INSTITUT FRANÇAIS DU PÉTROLE
ÉCOLE NATIONALE SUPÉRIEURE DU PÉTROLE ET DES MOTEURS

Marcel LATIL
Research Engineer

with the assistance of

Charles BARDON
Senior Research Engineer

Jacques BURGER
Senior Research Engineer

Pierre SOURIEAU
Assistant to the Director

all the authors are on the staff of
Reservoir Engineering Research Unit
Institut Français du Pétrole

ENHANCED OIL RECOVERY

Translation from the French
by Paul ELLIS

1980

GULF PUBLISHING COMPANY • BOOK DIVISION • HOUSTON, LONDON, PARIS, TOKYO

ÉDITIONS TECHNIP • 27 RUE GINOUX • 75737 PARIS CEDEX 15 technip

This Edition, 1980
Gulf Publishing Company
Book Division
Houston, Texas

ISBN 0-87201-775-3
Library of Congress Catalog Card No. 79-56344

Printed in France
by Imprimerie Louis-Jean, 05002 Gap

PREFATORY NOTE

Specialists from the petroleum industry, the *Institut Français du Pétrole (IFP),* the *Ecole Nationale Supérieure du Pétrole et des Moteurs (ENSPM)* and universities teach at the Graduate Study Center for Drilling and Reservoir Engineering of *ENSPM.*

In conjunction with this teaching, they have written various books dealing with the different scientific and technical aspects of these petroleum operations.

The present book, which is part of the series of production courses written under the guidance of A. Houpeurt, is one of them.

Contents

2. WATER INJECTION

3. GAS INJECTION IN AN OIL RESERVOIR (IMMISCIBLE DISPLACEMENT)

4. MISCIBLE DRIVE

6. THERMAL RECOVERY METHODS

by J. BURGER and P. SOURIEAU

7. OTHER METHODS OF ENHANCED RECOVERY

by C. BARDON and M. LATIL

REFERENCES

In addition to the references given at the end of each chapter, the two principal sources consulted during the preparation of this course were :

Munier-Jolain, J.P., Production – Tome VI – *Récupération Secondaire*. ENSPM course ; Editions Technip, July 1959.

Smith, C.R., *Secondary Recovery Mechanics*. Reinhold Publishing, July 1959.

NOTA : AIME notation has been used.

The references are given by chapter: Ref. 1 = reference 1 for that particular chapter.

introduction

The methods of calculation of reserves in place and of the recovery factors which could be expected by natural depletion were explained in Volume V of this production course. It was shown that the presence of a gas-cap or an active aquifer generally results in a high recovery factor, by providing a strong natural drive.

The lack of sufficient natural drive in most reservoirs has led operators to introduce some form of artificial drive, the most basic method being the injection of natural gas or water.

In the early days of the petroleum industry, reservoirs were allowed to produce naturally until a certain stage of depletion had been reached, generally when the production rates had become uneconomic. This was known as the "primary" production phase. In the second phase the recovery was increased by installing methods of artificial drive (water or gas injection) and these were logically known as "secondary recovery methods".

This definition of "secondary recovery" is only of historic interest, since secondary recovery methods are now introduced much earlier in the life of a field, often well before the end of the "primary" production phase.

However, before undertaking a "secondary" recovery project it should be clearly proven that the natural recovery processes are insufficient, otherwise there is a risk that the heavy capital investment required may be completely wasted. A certain amount of production data is therefore required. Before producing a reservoir it is unwise to assume that its associated aquifer or gas-cap is inactive. However, if a reservoir is produced too long during the primary phase the chances of a successful secondary recovery phase may be reduced. For example, if solution gas drive continues until a high gas saturation exists in the pores the final recovery by water injection may be low.

Several recovery methods may be used in succession : for example, initial recovery by natural drainage followed by water injection and then carbon dioxide injection. Some authors have used the term "tertiary recovery" to cover this situation.

The terms "secondary" and "tertiary" recovery thus describe the order in which the methods are used rather than being related to their characteristics. The authors of this present work prefer to avoid such terms and suggest a more explicit terminology.

Natural production depends on a reservoir's internal energy, and arises due to the existence of a higher pressure in the rock pores than at the bottom of the well. **All other recovery methods depend on the provision of additional energy** to improve the recovery of the remaining reserves.

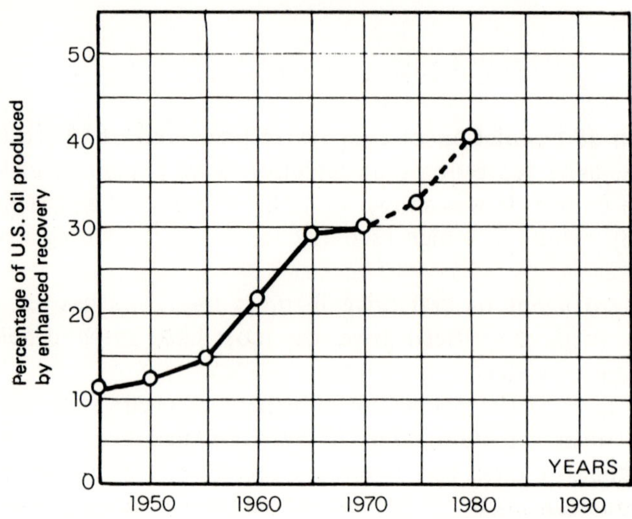

The importance of enhanced recovery in the USA (Data to 1970 and forecasts from S.W. Nicksic : Energy Sources, Vol. 1, No. 2 – Crane, Russak and Cy., Inc.)

Most methods provide the extra energy in mechanical form, by the injection of a fluid which displaces those already in place. This artificial sweep occurs under almost isothermal conditions.

There are also recovery methods in which only a small part of the energy supplied to the reservoir is mechanical. In the case of thermal recovery methods the fluid injected (mechanical energy) is capable of directly or indirectly supplying thermal energy to the reservoir. The thermal energy may be latent, as for example in the case of in-situ combustion, when the heat is generated by the reaction of oxygen in the injected air with part of the oil in place. The interaction between displacing and displaced fluids is highly affected by the temperature variations.

For the purpose of this volume all recovery methods other than natural production will be termed **enhanced recovery methods.**

Enhanced recovery plays a progressively more important part in oil production. Faced with ever-increasing demand and the continuing rise in oil prices, assisted recovery oil is becoming more and more competitive. **The diagram above** shows the situation in the United States (the forecasts were made before the 1973 oil crisis).

1

factors common to all enhanced recovery methods

11. PRINCIPAL INFLUENCES ON THE EFFICIENCY OF ENHANCED RECOVERY

The efficiency of an enhanced recovery method is a measure of its ability to provide greater hydrocarbon recovery than by natural depletion, at an economically attractive production rate.

In the **diagram below** we compare for a hypothetical reservoir three forecasts of cumulative hydrocarbon production as a function of time : $C(t)$. One curve is the forecast for natural depletion and the others are for two different proposed enhanced recovery methods, A and B[1]. All the forecasts start from the present time, T, up to which production has been obtained by natural depletion.

Any comparison should take into account not only $C(t)$ but also $\frac{dC(t)}{dt}$, the production rate. When $\frac{dC(t)}{dt}$ falls below a certain economic limit, enhanced recovery by the method in use must be discontinued.

The theoretical maximum recovery for any method (e.g. the asymptotes R_1, R_2 and R_3 (Fig. 11.1)) is only of academic interest, since it would only be approached at an uneconomic production rate.

The efficiency of an enhanced recovery method depends on:

(a) The reservoir characteristics.

(b) The nature of the displacing and displaced fluids.

---

(1) The importance of these curves was pointed out in Volume V of this ENSPM series. The forecasts are made by material balance or by empirical methods.

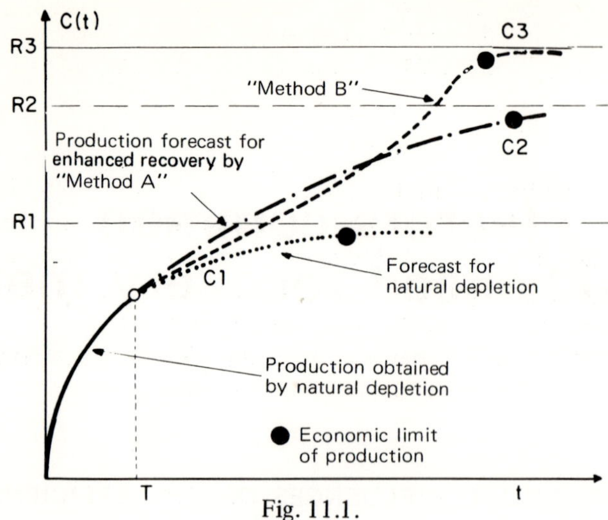

Fig. 11.1.

(c) The arrangement of production and injection wells (see sections 14 and 15).

11.1. The influence of reservoir characteristics

The following are some of the most important characteristics of a reservoir:

(a) Average depth.
(b) Structure, in particular the dip of the bed.
(c) Degree of homogeneity.
(d) Petrophysical properties (permeability, capillary pressure, wettability).

A. Depth

Reservoir depth has an important influence on both the technical and economic aspects of an enhanced recovery project.

On the technical level, a shallow reservoir puts a restraint on the injection pressure that can be used, since this must be less than fracture pressure.

Economically, the cost of an enhanced recovery project is directly related to depth, as reflected, for example, in the cost of drilling the extra wells required, or in the compressor power required in the case of gas injection.

B. Dip

For two phase flow in a inclined bed in which gravity opposes the sweep, the fractional flow (f_1) of the displacing fluid is given by:

$$f_1 = \frac{1 - \dfrac{k\,k_{r2}}{\mu_2\,u}\,\Delta\rho\,g\sin\alpha}{1 + \dfrac{k_{r2}\,\mu_1}{\mu_2\,k_{r1}}} < \frac{1}{1 + \dfrac{k_{r2}\,\mu_1}{\mu_2\,k_{r1}}}$$

for each value of the fluid saturation, S_1.

Using this equation it can be shown that the desaturation of (i.e. hydrocarbon recovery from) a porous medium, for any given economic limit of the fractional flow of the displacing fluid in the production stream, is greater when gravity plays a part (case B of Fig. 11.11) than when it does not.

In practice, gravitational forces are only truly effective in reservoirs containing highly permeable sands or in which the dip is unusually large.

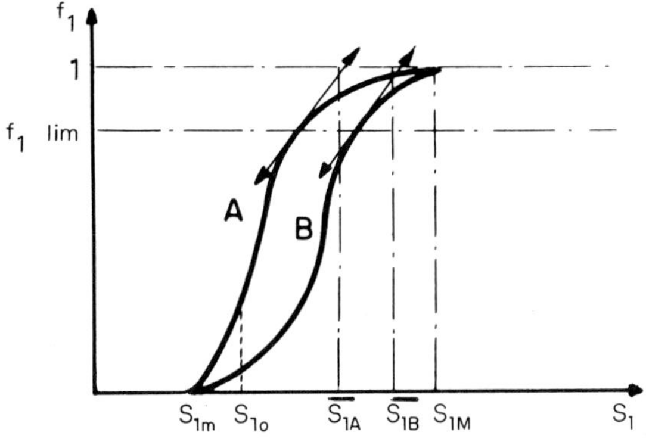

Fig. 11.11.

Moreover, if the transition zone is neglected (i.e. if we assume that the injected and displaced fluids are separated by a surface or "front" without thickness) the displacement front becomes unstable for injected fluid velocities higher than:

$$V_c = \frac{\Delta\rho\,gk\sin\alpha}{\phi_u\left(\dfrac{\mu_2}{k_{r2}} - \dfrac{\mu_1}{k_{r1}}\right)}$$

where

ϕ_u is the "useful" porosity; $\phi_u = \phi(S_{1M} - S_{1m})$; S_{1M} and S_{1m} are the limits of the saturation of the porous medium in fluid 1.

For horizontal beds([1]) the critical velocity is zero, injected water forming a tongue at the base of the bed and injected gas forming an umbrella at the top of

([1]) Also for down-dip injection of gas or up-dip injection of water.

the bed (Fig. 11.12). These phenomena cause rapid breakthrough of the injected fluid at the production wells.

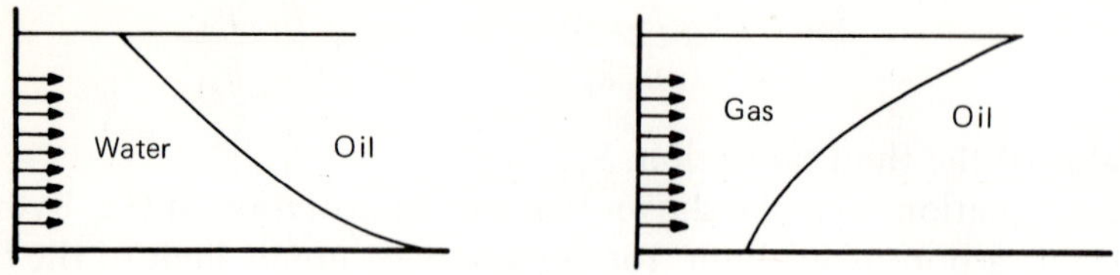

Fig. 11.12.

C. Homogeneity

In order to achieve a high recovery of hydrocarbons, there should be no impediment to fluid flow within the reservoir. Possible impediments may be of tectonic (e.g. isolating faults) or stratigraphic nature (e.g. lateral facies variation, lenses, unconformities). It is advisable to ascertain the degree of communication between wells before any enhanced recovery project, and to this end interference tests and well pressure histories are valuable.

In faulted and fissured reservoirs, and those with high permeability streaks, channeling allows the displacing fluid to bypass some of the oil in place and leads to a low recovery factor.

1. Layered reservoirs (Refs. 2 to 6)

Let us consider the case of a reservoir consisting of several non-communicating layers, initially containing a hydrocarbon B and undergoing a linear sweep by a non-miscible fluid A (Fig. 11.13).

Let us assume, for simplicity, that the layers are identical except for their thickness and permeability.

At the start of the displacement (only fluid B in place in the reservoir) the rate of injection Q_t of fluid A can be divided amongst the layers using the equation:

$$Q_i = Q_t \frac{k_i h_i}{\Sigma\, k_i h_i}$$

Thus in each layer a front separating the two fluids is formed which advances initially with a velocity proportional to k_i. The fronts are therefore staggered, being furthest ahead in the highest permeability layers.

If, as is very often the case, the injected fluid viscosity is less than that of the fluid in place, the differences between the front locations increase with time.

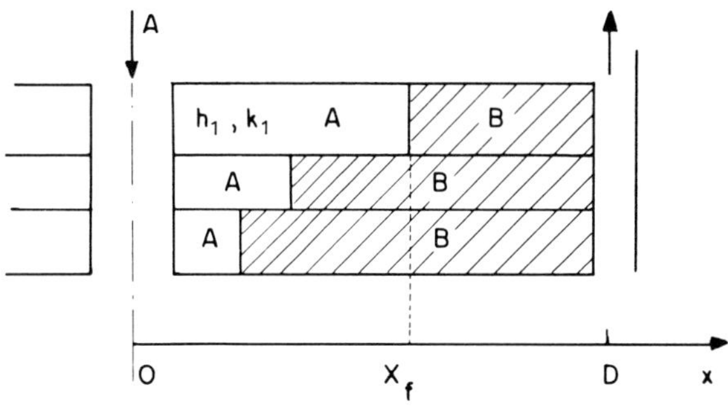

Fig. 11.13.

Injected fluid reaches the production wells via the most permeable layers, at which time the less permeable layers have been only partially swept.

Injection is normally discontinued when the percentage of injected fluid in the production reaches the economic limit. At this time the less permeable layers still contain considerable amounts of hydrocarbon in place.

To overcome these problems one may try to temporarily plug off the most permeable layers during part of the injection phase. Among practical methods are the injection of plugs of resin or cement and the use of wireline equipment (e.g. sliding sleeves, sidepocket mandrels) installed during the completion of the well.

This method of injection, known as selective injection, is only used in certain types of layered reservoir.

It should be noted that in practice we can never be certain that the layers are completely non-communicating, the quality of the seal between zones often being variable throughout the reservoir.

2. Reservoirs with random heterogeneities

In a reservoir consisting of a permeable medium interspersed with thin dense lenticular deposits, more or less parallel to the dip, the average horizontal permeability may be high (permeabilities in parallel) while the vertical permeability is low (permeabilities in series). In the extreme, the vertical segregation of fluids is negligible. However, the flood front does not advance uniformly throughout the reservoir, since a variety of fluid paths of varying rapidity are offered to the injected fluid. The first breakthrough occurs at an early stage of the displacement, breakthrough on other streamlines occuring later according to the permeability of the beds encountered.

D. Petrophysical properties

Porosity, permeability, relative permeability (as a function of saturation), capillary pressure and wettability are all properties which should be taken into account in the study of an enhanced recovery project.

The higher the porosity and the higher the residual oil saturation at the end of the natural recovery phase, the more attractive an enhanced recovery project becomes.

For enhanced recovery as for natural recovery a high permeability is encouraging (high initial oil saturation, larger pore throats, etc.). However, the higher the permeability the greater the chance that the natural recovery will be so high that any enhanced recovery project would be uneconomic.

The permeability distribution in the reservoir depends on the degree of homogeneity and its effect has already been discussed.

The effect of capillary forces on recovery efficiency depends on the rate of production. It is occasionally beneficial, as for example when it helps to maintain a uniform front between two immiscible fluids in a heterogeneous porous medium (imbibition). But capillary forces often have a detrimental effect, being responsible for the trapping of oil within the pores. This trapping is a function of the ratio $\dfrac{V\mu}{\sigma \cos \theta}$ (ratio of viscous forces to capillary forces), V being the velocity of the front.

According to the special core analyses performed by Moore and Slobod (Ref. 9), the residual oil saturation decreases as the ratio $\dfrac{V\mu}{\sigma \cos \theta}$ increases.

The essential role of relative permeability was presented in Volume IV.

11.2. The influence of fluid characteristics

The principal fluid property to be taken into account when designing an enhanced recovery project is the viscosity.

If the fluids are highly viscous the displacement velocities will be low, since the applied pressure gradients are limited. Oil production will be at such a low rate that it will not be economically attractive.

For a given volume of injected fluid, all other things being equal, the residual oil saturation will be higher the higher the oil viscosity.

Let us re-examine the fractional flow equation for the case of constant capillary pressure in the porous medium (Ref. 8):

$$f_1 = \frac{1 - \dfrac{k\,k_{r2}}{\mu_2 u}\,\Delta\rho\, g \sin\alpha}{1 + \dfrac{k_{r2}\,\mu_1}{k_{r1}\,\mu_2}}$$

differentiating with respect to μ_2:

$$\frac{\partial f_1}{\partial \mu_2} = \frac{1 - f_1}{k_{r2}\left(\dfrac{\mu_1}{k_{r1}} + \dfrac{\mu_2}{k_{r2}}\right)}$$

As f_1 is always less than 1, $\partial f_1/\partial \mu_2$ is always positive. Displacing fluid fractional flow curves for various oil viscosities are shown on Fig. 11.21 for Bartlesville sandstone swept with water.

For all the oils used in this experiment, the final oil recovery (injection continued until only water was produced: $f_1 = f_w = 1$) is the same: $(0.80 - 0.20)/0.80$ or 75 % of the oil originally in place.

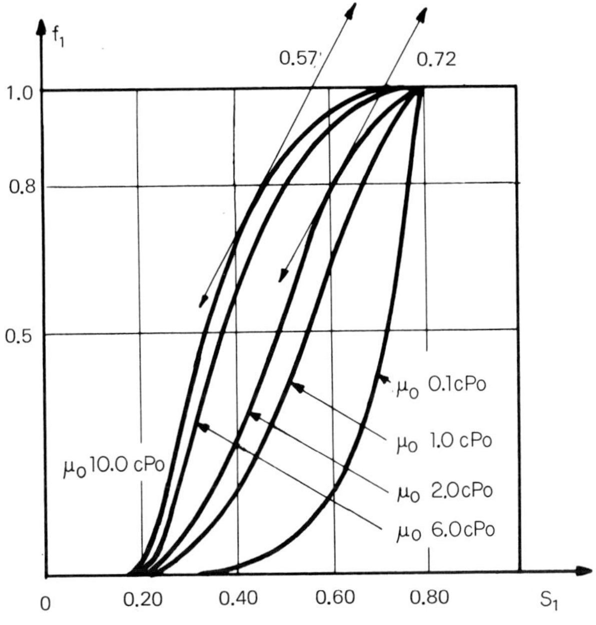

Fig. 11.21.
From Pirson (Ref. 7)

However, for economic reasons the wells would be abandoned at a certain limiting water cut, and it can be seen that, at any given value of f_1 lesser than one, the recovery decreases as the oil viscosity increases.

Thus, for a fractional flow of 0.8, oil recovery obtained from Fig. 11.21 is as follows:

$$2 \text{ cP oil} \quad r = \frac{0.72 - 0.2}{0.8} = 0.65$$

$$10 \text{ cP oil} \quad r = \frac{0.57 - 0.2}{0.8} = 0.46$$

Viscosity has a further important effect on sweep efficiency in that it is one of the parameters which determine the mobility ratio. This will be discussed in the next chapter.

12. LINEAR DISPLACEMENT

A linear displacement is one in which the fluid velocities have a constant direction at every point and for all time. It is the limiting case of a number of displacements (gas-cap expansion, bottom water-drive, gas injection into a gas-cap, water injection into an aquifer, etc.). The study of linear displacements can be divided into two parts:

(a) The theory of frontal displacement.
(b) The theory of piston-like displacement.

12.1. Frontal displacement

The theory of frontal displacement was proposed by Buckley and Leverett and expanded by Welge. It has been studied in detail in Volume IV of the production course. We shall only summarise the principal results here.

According to the theory, injection of a fluid 1 in a porous medium initially containing fluid 1 at a uniform saturation S_{1i} and fluid 2 at a uniform saturation $1 - S_{1i}$, leads to the formation of a front which advances at a velocity:

$$V = \frac{Q_T}{A\phi}\left(\frac{\mathrm{d}f_1}{\mathrm{d}S_1}\right)_{S_1 = S_{1F}}$$

where

S_{1F} is the saturation in fluid 1 immediately behind the front,
Q_T is the injection rate (assumed constant),
A is the cross-sectional area.

In the zone swept by the injected fluid 1, fluids 1 and 2 are both flowing. By the definition of fractional flow, in a section of the porous medium in which the saturation in fluid 1 is S_1:

(a) The flow of fluid 1 is $Q_1 = f_1 Q_T$.
(b) The flow of fluid 2 is $Q_2 = (1 - f_1) Q_T$.

Neglecting capillary forces the fractional flow f_1 is given by:

$$f_1 = \frac{1 - \dfrac{Ak\,k_{r2}}{Q_T \mu_2} \Delta\rho\, g \sin\alpha}{1 + \dfrac{k_{r2}}{k_{r1}} \dfrac{\mu_1}{\mu_2}} \quad \text{(Eq. 12.11)}$$

where
α is the angle between the horizontal and the direction of flow,
$\Delta\rho = \rho_1 - \rho_2$.

In the case of a horizontal reservoir being either very thin or subject to high flow rates, the equation for f_1 reduces to:

$$f_1 = \frac{1}{1 + \dfrac{k_{r2}}{k_{r1}} \dfrac{\mu_1}{\mu_2}} \quad \text{(Eq. 12.12)}$$

Note

For a relatively homogeneous reservoir the application of Eq. 12.11 requires the knowledge of the two relative permeability functions $k_{r1}(S_1)$ and $k_{r2}(S_1)$. It is necessary to perform laboratory experiments on a large number of representative samples, from which an average curve for each of the two relative permeabilities can be obtained. However, the application of Eq. 12.12 requires only the knowledge of the ratio $\dfrac{k_{r2}}{k_{r1}}(S_1)$.

This may be obtained in one of three ways:

(a) By using laboratory measurements as before.

(b) By measuring producing water-oil or gas-oil ratios during the natural depletion phase, calculating the average saturations in the reservoir at the same time by material balance. In this case the curve obtained will need extrapolation.

(c) By observing the results of the displacement of fluid 2 by fluid 1 at the first line of production wells, during the first phase of an enhanced recovery project.

Having plotted f_1 vs. S_1, the saturation at the front, S_{1F}, is determined by drawing the secant line from the point on the curve representing the initial fluid saturation S_{1i}, to the higher saturation S_{1F} at which the secant is a tangent to the curve (Fig. 12.11).

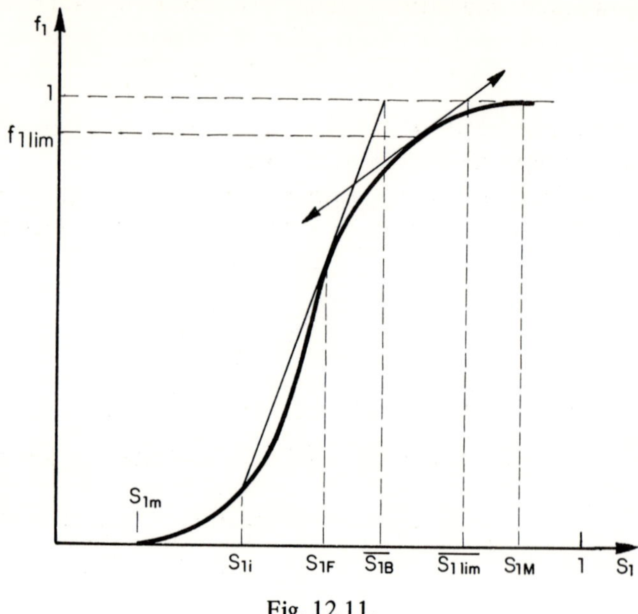

Fig. 12.11.

We may then calculate the time of the arrival of the front at the production well, taking the time at the start of injection of fluid 1 to be zero. If L is the distance between the injection and production wells, the breakthrough time is given by:

$$t_B = \frac{L}{\frac{Q_T}{A}\left(\frac{\mathrm{d}f_1}{\mathrm{d}S_1}\right)_{S_1=S_{1F}}}$$

Conversely, when relative permeability curves are not available, the observation of the discontinuity of producing water/oil ratio (WOR) or gas-oil ratio (GOR) at breakthrough allows us to calculate the saturation S_{1F} in the vicinity of the wellbore at this time:

$$S_{1F} - S_{1i} = \frac{C_I}{V_p}\left[\left(\frac{Q_1}{Q_T}\right)_B - \left(\frac{Q_1}{Q_T}\right)_i\right]$$

where

V_p is the pore volume,

C_I is the cumulative volume injected at breakthrough, and

the subscript i refers to the initial conditions.

Normally injection is continued after breakthrough until a limiting fraction of fluid 1 is obtained in the production, this fraction being $f_{1\,\mathrm{lim}}$. The corresponding average saturation in the porous medium is $\overline{S_{1\,\mathrm{lim}}}$.

For linear flow, such as occurs in laboratory core samples, the oil recovery is given by:

$$E_d = (\overline{S_{1\,\text{lim}}} - S_{1i})/(1 - S_{1i})$$

E_d represents the replacement of the fluid in place by the injected fluid and is known as the **microscopic displacement efficiency.**

In the case of a field with several rows of wells, the saturation in the vicinity of the first line of production wells may be calculated by integrating the saturations from 0 to L:

$$S_{1L} - S_{10} = \frac{C_I(t)}{V_p}\left(1 - \frac{Q_2(t)}{Q_T}\right) - \frac{C_1(t)}{V_p}$$

where $C_1(t)$ is the cumulative production of fluid 1 at time t.

If a fraction, s, of the injected fluid is withdrawn at the first line of producers:

$$S_{1L} - S_{10} = \frac{C_I(t)}{V_p}\left(1 - \frac{Q_{2P}(t)}{sQ_T}\right) - \frac{C_{1P}(t)}{sQ_p} \qquad \text{(Eq. 12.13)}$$

where

$Q_{2P}(t)$ is the rate of flow of fluid 2 at the first line of production wells at time t, and

$C_{1P}(t)$ is the cumulative production of fluid 1 at the first line of production wells at time t.

We are assuming here that the productions of fluids 1 and 2 at the wells are in proportion to the amounts of each fluid flowing in the formation near the wellbores.

Equation 12.13 enables us to calculate the theoretical curves $f(S)$ using the records of production from the first line of wells. It is then possible to make production forecasts for the other lines of wells.

A generalised form of the theory of frontal displacement is given in the appendix.

12.2. Piston-like displacement

The theory of piston-like displacement is an extreme simplification of the problem of the displacement of one fluid by another. It assumes that there is no displaced fluid movement behind the front. In practical terms this implies that the oil saturation behind the front is at its residual value, S_{or}, and that the microscopic displacement efficiency $E_d = (1 - S_{1i} - S_{or})/(1 - S_{1i})$.

Figure 12.21 summarises the differences between the saturation profiles calculated by the Buckley-Leverett theory and that of piston-like displacement.

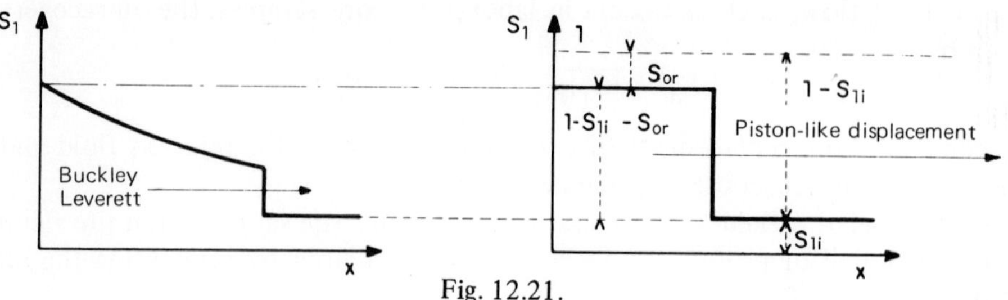

Fig. 12.21.

The theory of piston-like displacement is without doubt a gross simplification in the case of a reservoir subjected to a pseudo-linear sweep, since it is certain that just behind the front a two-phase flow region exists. However, this region is often of limited extent, and its influence becomes negligible when it represents less than 5 % of the swept pore volume. In particular it may be ignored when the distances between injection and production wells are large, or when the oil in place is viscous.

Let us consider the linear piston-like displacement of a fluid 2 of mobility $\lambda_2 = (k_2/\mu_2)$ by a fluid 1 of mobility $\lambda_1 = (k_1/\mu_1)$, corresponding to a mobility ratio $M = \lambda_1/\lambda_2$, through a porous medium of length X (Fig. 12.22). The pressure difference Δp between the entrance and exit faces of the medium is held constant. If M is not unity, we would expect that the velocity of the front would be a function of its position.

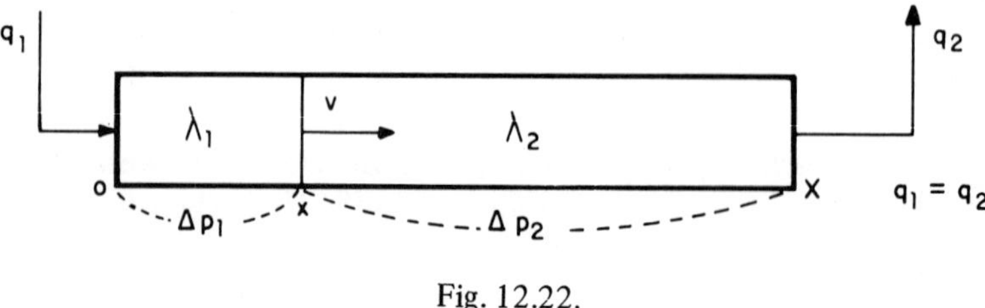

Fig. 12.22.

Consider the time at which the front is located at a distance x from the entrance face. The velocities of the fluids on either side of the front are identical, thus:

$$\left(\frac{\mathrm{d}p}{\mathrm{d}x}\right)_2 = M\left(\frac{\mathrm{d}p}{\mathrm{d}x}\right)_1$$

If the pressure losses in the zones behind and ahead of the front are Δp_1 and Δp_2, we have:

$$\Delta p = \Delta p_1 + \Delta p_2$$

$$\Delta p_1 = x\left(\frac{dp}{dx}\right)_1$$

$$\Delta p_2 = (X - x)\left(\frac{dp}{dx}\right)_2$$

Solving for $\left(\frac{dp}{dx}\right)_1$ we find:

$$\left(\frac{dp}{dx}\right)_1 = \frac{\Delta p}{MX + x(1 - M)}$$

From Darcy's law:

$$v = \frac{dx}{dt} = \frac{u_1}{\phi_D} = -\frac{\lambda_1}{\phi_D}\left(\frac{dp}{dx}\right)_1 = \frac{-\lambda_1 \Delta p}{\phi_D [MX + x(1 - M)]}$$

where

$$\phi_D = \phi(S_{1M} - S_{1m})$$

By integrating we can find the time taken for the front to travel from 0 to x:

$$t = \frac{\phi_D \left(MXx + \frac{x^2}{2}(1 - M)\right)}{\lambda_1 \Delta p} \qquad \text{(Eq.12.21)}$$

Solving for x:

$$x = \frac{-MX + \sqrt{(MX)^2 - 2\lambda_1 \Delta p\, t[(1 - M)/\phi_D]}}{1 - M} \qquad \text{(Eq.12.22)}$$

If v_0 is the initial velocity of the front (velocity at $x = 0$) we have:

$$\frac{v}{v_0} = \frac{1}{1 + \frac{x}{X}\left(\frac{1}{M} - 1\right)}$$

Figure 12.23 shows how this velocity ratio varies with the position of the front for different values of M.

If M is greater than 1, the velocity of the front increases with x, and vice-versa. Although these calculations are based on a linear displacement, this result also holds true for other geometries.

It is convenient in frontal advance theory to introduce the terms conductance and conductance ratio.

The conductance is the ratio $\dfrac{q}{\Delta p} = \dfrac{-\lambda_1 A}{MX + x(1 - M)}$.

The conductance ratio γ is the ratio of the conductance of the system when the front is at x to the initial conductance of the system ($x = 0$). For linear displacement the equation is:

$$\gamma = \frac{\left(\frac{q}{\Delta p}\right)_x}{\left(\frac{q}{\Delta p}\right)_0} = \frac{M}{M + \frac{x}{X}(1 - M)}$$

The variation of γ with x for different mobility ratios is shown in Fig. 12.24.

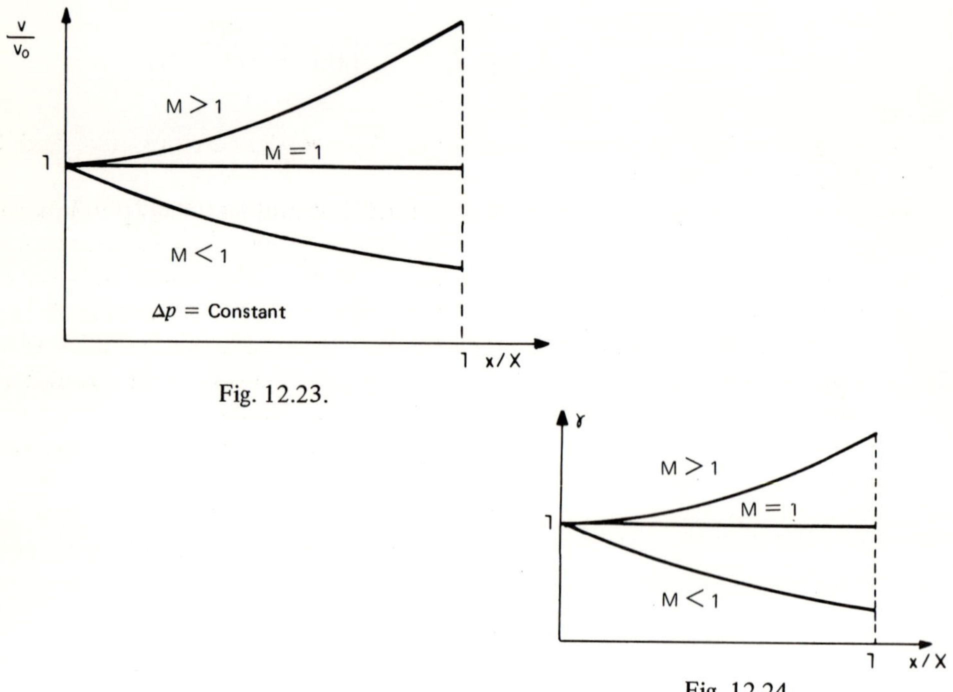

Fig. 12.23.

Fig. 12.24.

13. TWO AND THREE-DIMENSIONAL DISPLACEMENT

In order to study multi-dimensional muti-phase flow problems it is often necessary to use numerical simulation by computer. The calculations are fairly involved, relatively time-consuming and therefore quite expensive. It is possible, however, to do without them, at least for a first approximation.

Consider an injection well I surrounded by three production wells A, B and C. (Fig. 13.1).

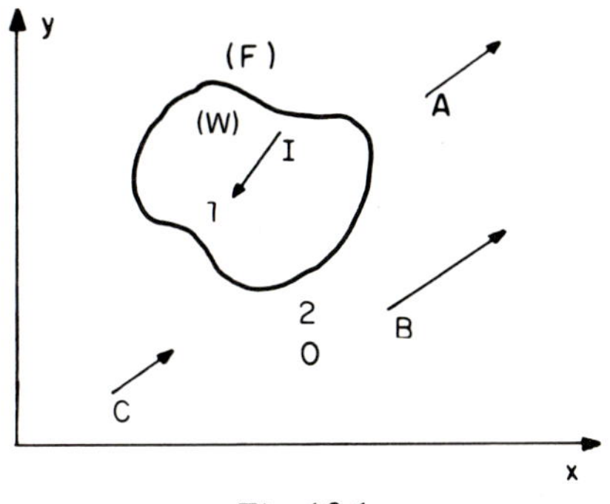

Fig. 13.1.

At all times a front F separates the region W invaded by the injected fluid 1 from the uninvaded region O.

Let us assume that in the region O only fluid 2 of mobility k_{r2}/μ_2 is flowing. In region W there is simultaneous flow of the two fluids. Along a streamtube invaded by fluid 1 the saturation S_1 decreases regularly from upstream (injection well) to downstream (Buckley-Leverett front). If the geometry of the streamtube remains constant as the displacement progresses, we can substitute an average uniform saturation $\overline{S_1}$ for the saturation profile; this saturation being consistent with the material balance and the frontal velocity.

In order to study the advance of the front within the reservoir, the average conductivity of the fluids in zone W must be defined. It is usual to attribute the sum of the conductivities of fluids 1 and 2 behind the front to phase 1, in such a way as to respect the pressure losses between the injection well and the front along a streamtube:

$$\int_0^{X_F} \frac{\mu_1 Q_T}{k\,\overline{k_{r1}}} \frac{dx}{A(x)} = \int_0^{X_F} \frac{Q_T}{k} \left[\frac{\mu_1 f_1}{k_{r1}(S_1)} + \frac{\mu_2(1-f_1)}{k_{r2}(S_2)} \right] \frac{dx}{A(x)}$$

Craig established during studies with petrophysical models that $\overline{k_{r1}}$ was close to the permeability $k_{r1}(\overline{S_1})$. It is wise, however, to calculate $\overline{k_{r1}}$ by using the above formula for the linear case ($A(x)$ constant) (Fig. 13.2).

The mobility ratio of the fluids in the regions W and O is given by the equation:

$$M = \frac{\overline{k_{r1}}\,\mu_2}{\mu_1 k_{r2}(S_{1i})} \approx \frac{k_{r1}(\overline{S_1})\,\mu_2}{\mu_1 k_{r2}(S_{1i})}$$

In this book M will always denote the ratio of the mobility of the displacing fluid to that of the displaced fluid.

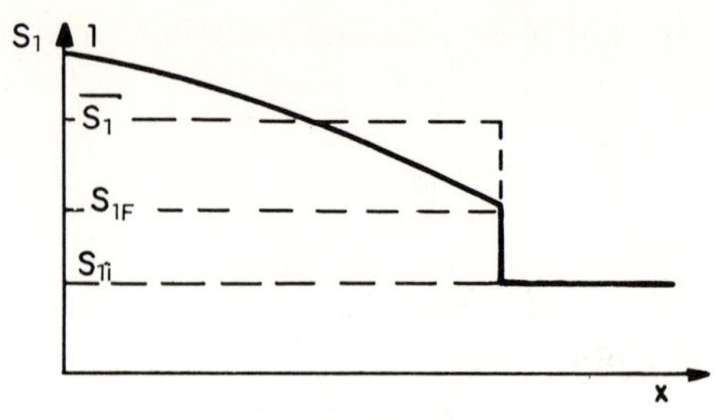

Fig. 13.2.

The area of the region W increases with time, but never becomes equal to the area of the reservoir, even long after the initial breakthrough of the front at the production wells. There always remain certain "dead zones" not swept by the displacing fluid. The ratio E_s = Area W/Reservoir Area is called the superficial or **areal sweep efficiency**.

From this areal sweep efficiency we can calculate a volumetric sweep efficiency E_v by weighting E_s by $h\phi S$ (h = reservoir thickness, ϕ = porosity, S = gas or oil saturation). In the case of constant thickness, porosity and saturation we have $E_v = E_s$.

If there are vertical heterogeneities within the volume swept, some zones will not be invaded by the displacing fluid. This can be accounted for by using an invasion efficiency E_i of less than unity. E_i is also known as the vertical sweep efficiency.

The overall sweep efficiency is defined as the product:

$$E = E_s \times E_d \times E_i$$

where

E_d is the previously defined microscopic displacement efficiency.

• Description of a theoretical displacement

Let us consider an injection well I surrounded by 4 production wells A, B, C and D in a homogeneous isotropic square reservoir of constant thickness. We shall assume that the injection pressure is constant and that the flowing pressures at the four production wells are equal and constant.

At the start of injection, the front takes on a practically cylindrical form around the injection well (Fig. 13.3a).

However, the various streamlines between the injection well and a production well (shown dashed in Fig. 13.3) have different lengths. There will therefore be a range of pressure gradients and hence velocities along the different streamlines.

The front gradually deforms until at breakthrough it has the shape shown in Fig. 13.3b.

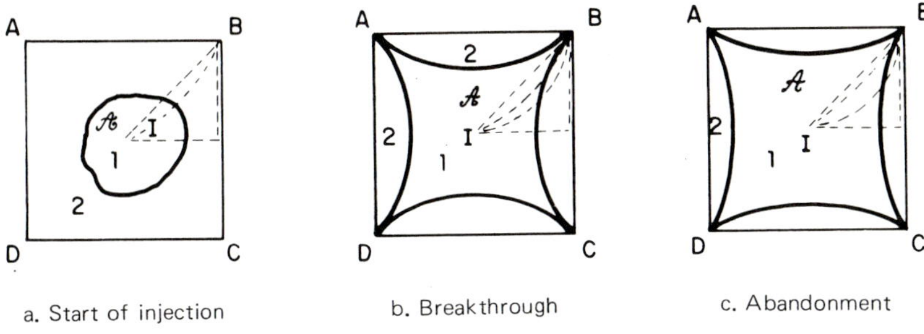

Fig. 13.3.

If injection is continued after breakthrough until abandonment, the area $\mathscr{A}$ invaded by fluid 1 continues to increase but always remains less than that of the square. The areal sweep efficiency continually increases but never reaches unity. It should be noted that the oil produced comes partly from oil located in zones previously unswept and partly from the continually improving displacement efficiency in the zones already swept.

14. INJECTION WELL LOCATION

The relative location of injection and production wells depends on the geology of the reservoir, its type, and the volume of hydrocarbon-bearing rock required to be swept in a time limited by economics.

It is advantageous, where possible, to make use of any favourable influence of gravity, for example in inclined reservoirs, reservoirs with a gas-cap or with an underlying aquifer.

This leads to two types of injection well location:

(a) Central and peripheral flooding, in which the injectors are grouped together.

(b) Pattern flooding, in which the injectors are distributed amongst the production wells.

14.1. Central and peripheral flooding

This type of injection occurs in the following cases:

(a) Reservoir with a gas-cap in which gas injection is taking place. If the reservoir is a fairly regular anticlinal structure, the injection wells are normally grouped in a cluster around the top of the anticline (Fig. 14.11).

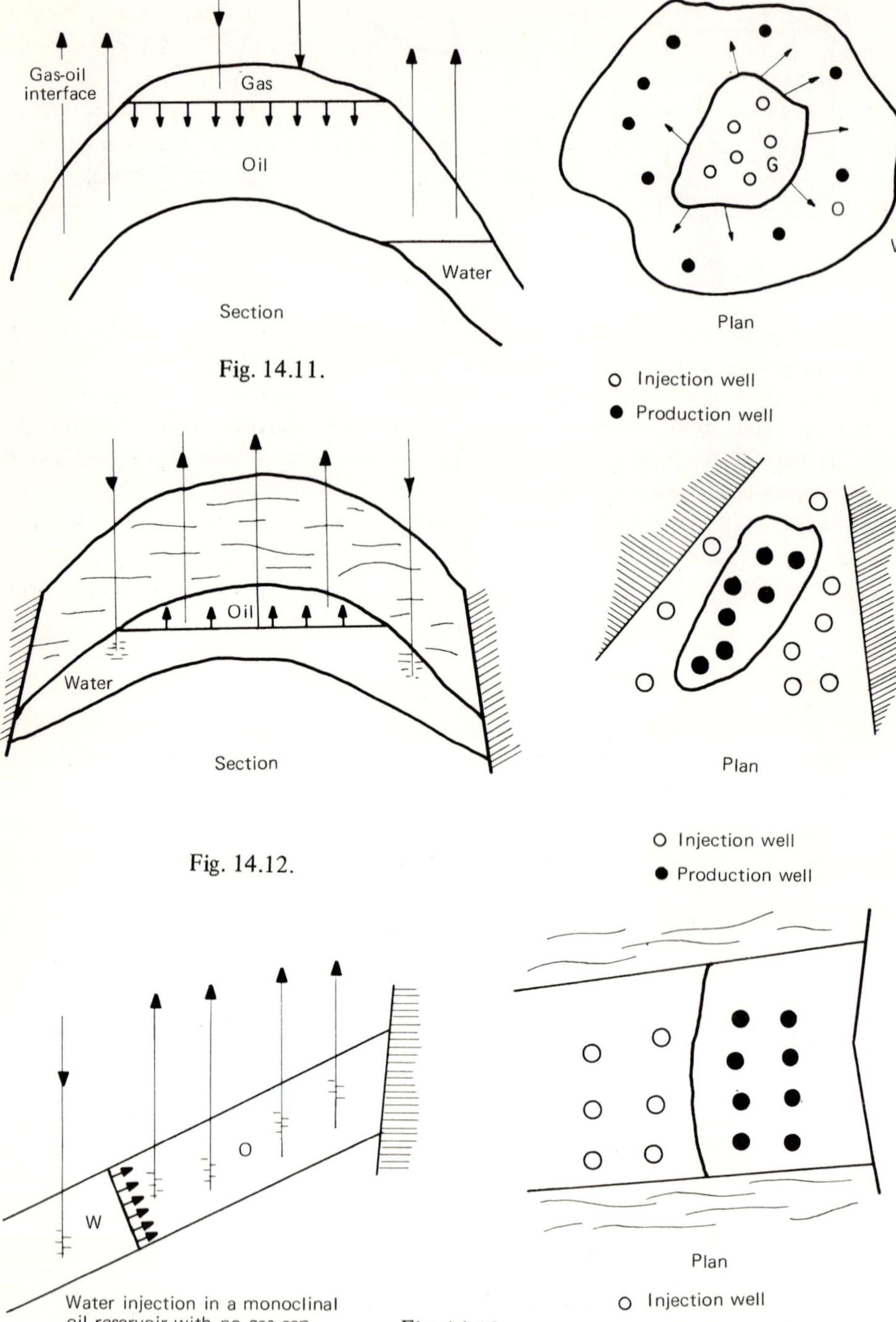

Fig. 14.11.

Fig. 14.12.

Fig. 14.13.

(b) Anticlinal reservoir with an underlying aquifer in which water injection is taking place. In this case the injectors will form a ring around the reservoir (Fig. 14.12).

(c) Monoclinal reservoir with gas-cap or aquifer undergoing gas or water injection. The injectors are grouped in one or more lines located towards the base of the reservoir in the case of water injection, towards the top in the case of gas injection (Fig. 14.13).

14.2. Pattern flooding

Pattern flooding is principally employed in reservoirs having a small dip and a large surface area. In order to ensure a uniform sweep the injection wells are distributed amongst the production wells. This is done either by converting existing production wells into injectors or by drilling infill injection wells. In both cases the aim is to obtain as uniform a distribution of wells as was used for the natural recovery phase.

Historically, due to the fact that the oil leases were divided into square miles and quarter square miles, *US* fields were developed in a very regular fashion. The various regular well patterns have been much documented and studied.

The most common patterns are the following:

A. *Direct line drive (Fig. 14.21)*

The lines of injection and production wells are directly opposed. The system is characterised by the two parameters:

a = spacing between wells of the same type,
d = spacing between lines of injection and production wells.

B. *Staggered line drive*

The wells are in lines as before, the injectors and producers being no longer directly opposed but laterally displaced, normally by a distance of $a/2$.

C. *Five-spot (Fig. 14.22)*

This is a particular case of staggered line drive in which $d/a = 1/2$, and is the most well-known pattern. Each injection well is located at the centre of a square defined by four production wells. In patterns A, B and C the injection and production wells are equal in number ($I/P = 1$).

D. *Seven-spot (Fig. 14.23)*

The injection wells are located at the corners of a hexagon with a production well at its centre. There are twice as many injection wells as production wells ($I/P = 2$).

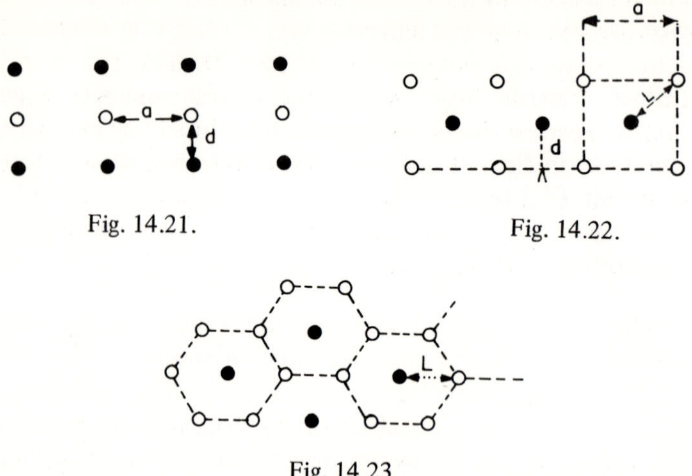

Fig. 14.21.

Fig. 14.22.

Fig. 14.23.

E. Nine-spot

The pattern is similar to that of a five-spot, but with an extra injection well drilled at the middle of each side of the square. There are three times as many injection wells as production wells ($I/P = 3$).

Notes

It is usual to name a regular pattern by the number of injection wells surrounding each producer plus one.

There are also inverted patterns in which the injection and production well locations are reversed with respect to the classical patterns:

Classical seven-spot $I/P = 2$.
Inverted seven-spot $I/P = 1/2$.

In practice, the choice of pattern is normally limited to either a line drive or a five-spot, since other patterns may require the drilling of additional wells.

The choice is directed in part by technical factors, but is principally constrained by economics. Indeed, the cost of drilling new injection wells represents the major part of the investment in an enhanced recovery project.

On the other hand, the conversion of producers into injectors reduces the production capacity of a field. Recovery will therefore be prolonged and profitability reduced. The choice between drilling new wells and converting old ones can be decided by economic analysis, provided that there are no technical constraints, in particular in respect of the conversion of old wells. It must first be ascertained that the old wells are suitable for conversion (that the tubing is in good condition, that there is no skin damage etc.).

It should be noted that in the case of an anisotropic reservoir a regular pattern will not necessarily be the most efficient.

Consider, for example, a field developed on a classical five-spot pattern. Let us assume that the permeability tensor k has its principal directions coincident with the x and y axes and that k_x is much higher than k_y (Fig. 14.24a).

We can see that the swept area at breakthrough is small. Assuming that, being aware of this phenomenon, we increase the well spacing in the x direction while maintaining the spacing in the y direction, we shall see an increase in the areal sweep efficiency at breakthrough (Fig. 14.24b).

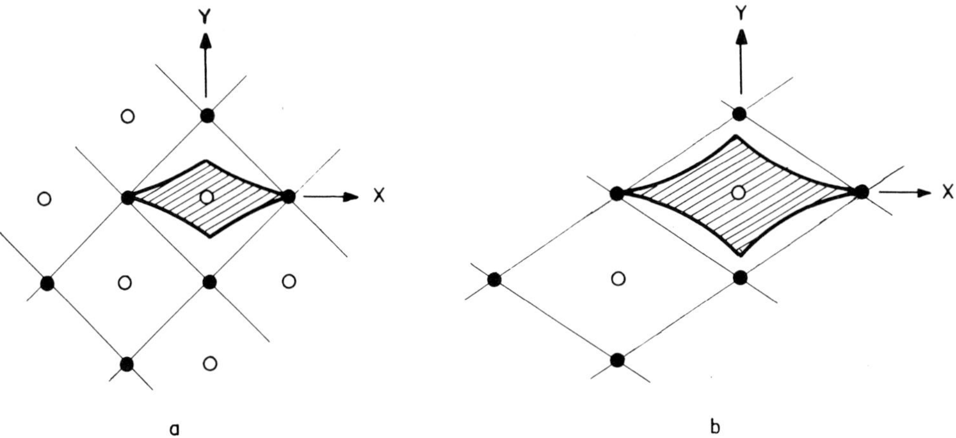

Fig. 14.24. Anisotropic reservoir $k_x > k_y$.
Swept area at breakthrough.

We usually under-estimate the effect of anisotropy, believing it to be much less important than that of heterogeneity. This fact is to be regretted, since there is no reason why k_x and k_y should not vary randomly with x and y while k_x/k_y remains constant. This type of system could then be treated as if it were isotropic, after a suitable adjustment of the shape of the pattern.

However, it should be pointed out that it is not easy to define the anisotropy of a reservoir (Ref.12). This requires, amongst other things, the availability of oriented cores.

15. AREAL SWEEP EFFICIENCY FOR PATTERN FLOODS

Research into the areal sweep efficiency of the various regular injection patterns has involved numerous theoretical and practical studies, the latter being mainly performed with petrophysical models.

It appears that the areal sweep efficiency always decreases as the mobility ratio M increases ($M = k_1\mu_2/k_2\mu_1$).

This phenomenon can be easily explained. Let us consider, for example, the five-spot pattern studied in Section 13, Fig. 13.3.

We saw that the unequal path lengths travelled by different fluid particles gave rise to a variety of pressure gradients and thus a variety of frontal displacement velocities for the different paths. In the case of practically incompressible fluids, or those whose density changes only slightly during the displacement (thus excluding any change of state), this general phenomenon is more pronounced when the mobility ratio is greater than 1 and progressively more so the greater M is. In this case, at any given time the front will have moved furthest along the shortest streamline (a), since the resistance to flow is lower along this streamline than along the others, the injected fluid 1 being more mobile than the fluid in place (Fig. 15.1). The inequality of velocity increases throughout the displacement and a markedly deformed front is obtained. Early breakthrough of the front at the production wells occurs and the resulting areal sweep efficiency is very poor. Figure 15.2 shows the difference in swept area obtained at breakthrough for extreme values of the mobility ratio.

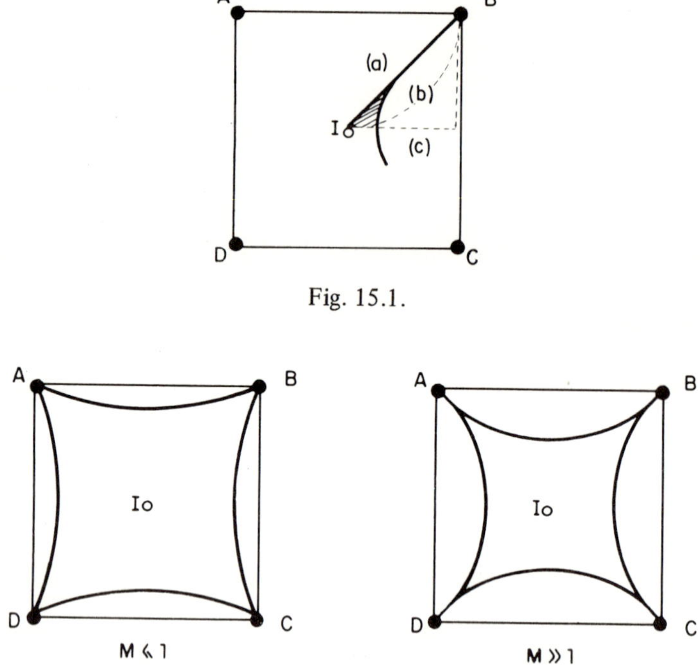

Fig. 15.1.

Fig. 15.2. Swept area at breakthrough as a function of mobility ratio for a five-spot in a homogeneous isotropic medium.

In theory, displacements at mobility ratios greater than 1, in practice greater than 2, suffer from instability known descriptively as fingering, which tends to further reduce the areal sweep efficiency (in the case of a hypothetical homogeneous and isotropic reservoir).

15.1. Unit mobility ratio

The first authors who attempted to quantify the areal sweep efficiency of the various patterns, amongst them Muskat, studied the case of two fluids of equal mobility, $M = 1$, in a homogeneous, horizontal reservoir of constant thickness.

Figure 15.11 shows the variation of E_s at breakthrough for direct and staggered line drives as a function of the ratio d/a, according to Muskat and Prats.

Muskat established mathematically (Ref. 10) that for a direct line drive with $d/a \geqslant 1.5$:

$$E_s = 1 - 0.441 \frac{a}{d}$$

For the staggered line drive, for which all authors have limited themselves to the case in which the alternate rows of wells are displaced by one-half of the interwell distance, the calculations published by Prats are regarded as the more rigorous.

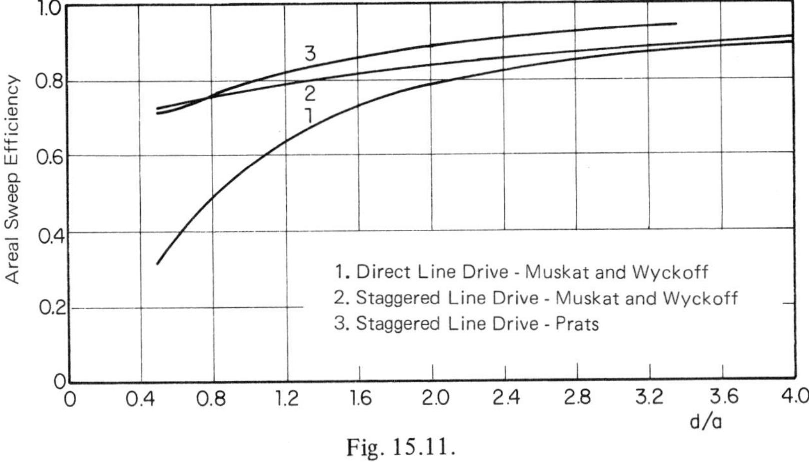

Fig. 15.11.

Muskat also studied the conductivity of the direct and staggered line drive systems, that is the ratio $\frac{q}{\Delta p}$: production rate per well per unit pressure difference between injection and production wells at steady-state conditions. The results obtained for various values of d/a are shown in Fig. 15.12 (Ref. 10).

For $d/a \geqslant 1$ the curves 1, 2 and 3 are defined by the same equation, valid in any consistent units system:

$$\frac{q}{\Delta p} = \frac{2kh/(\mu B)}{\dfrac{d}{a} - 1.17 + \dfrac{2}{\pi} \log \dfrac{a}{r_w}}$$

where

B is the formation volume factor and,

r_w is the radius of the wellbore.

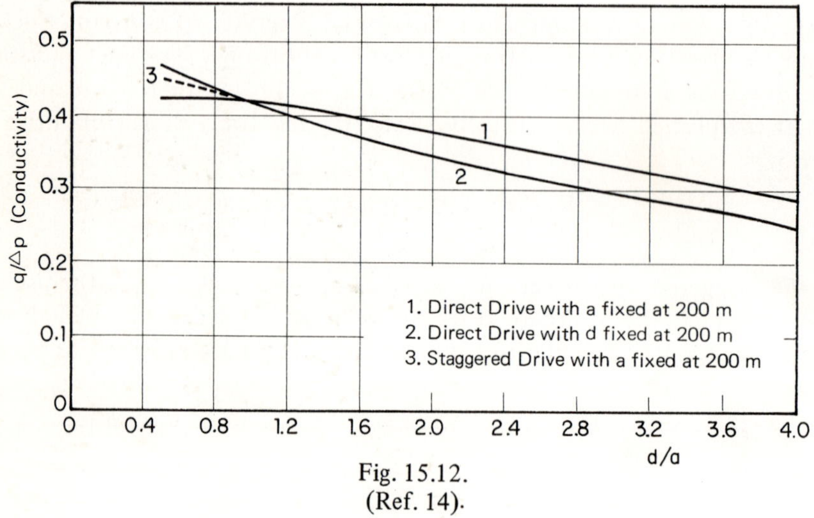

Fig. 15.12.
(Ref. 14).

We can see that the conductivity decreases as the system becomes more elongated. (Lengthening of the streamlines). The highest conductivity is obtained when $d/a = 0.5$.

It has been found that for a homogeneous, horizontal reservoir of constant thickness, the areal sweep efficiency E_s at breakthrough is around 70 % for a five-spot and 74.5 % for a seven-spot pattern.

15.2. Non-unit mobility ratio

Various authors, using different experimental or analytical techniques, have investigated the variation of areal sweep efficiency as a function of mobility ratio. Significant differences are to be found between their results.

We show here the types of curve obtained by Caudle and Witte for the five-spot pattern. Similar figures are available for other geometries.

Figure 15.21 shows E_s as a function of M for various values of f_1 (the fraction of displacing fluid in the production) and Fig. 15.22 shows E_s as a function of M for various values of $\frac{V_I}{V_D}$ (V_I = pore volume injected, V_D = displaceable pore volume). The displaceable pore volume is defined as the product of the pore volume and the maximum change in oil saturation.

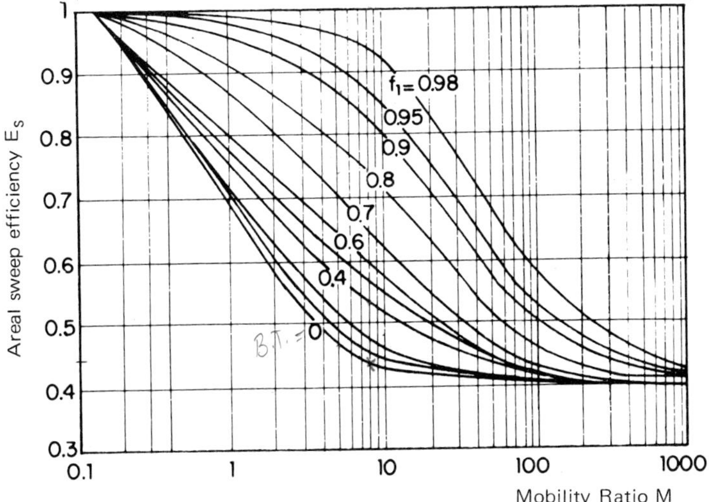

Fig. 15.21.
(Ref. 14).

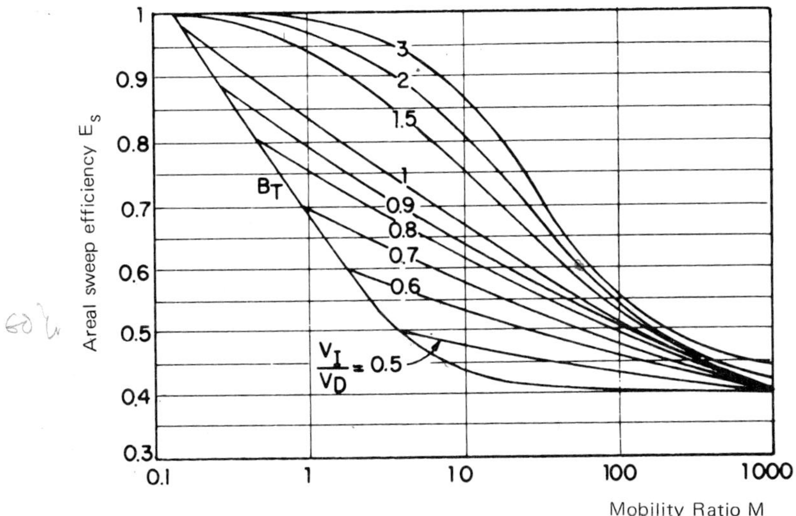

Fig. 15.22.
(Ref. 14).

The following example shows how Figs. 15.21 and 15.22 may be used.

Example 1: Recovery by water injection in a five-spot

Area = 12.5 km^2; thickness = 10 m; ϕ = 0.20; S_{oi} = 0.7; S_{or} = 0.3; μ_o = 5 cP; μ_w = 0.8 cP; k_{ro} = 0.63; k_{rw} = 0.21; injection rate = 5 000 m^3/day; B_o = 1.25; B_w = 1.0.

Solution

$$V_D = A\phi h(S_{oi} - S_{or}) = 12.5 \times 0.2 \times 10 \times 0.4 \times 10^6 = 10^7 \text{ m}^3$$

$$\text{Mobility Ratio } M = \frac{0.21 \times 5}{0.63 \times 0.8} = 2.08$$

For M = 2.08, we read from Fig. 15.21 that $E_s \approx 0.6$ at breakthrough ($f_1 = 0$).

Oil recovery at breakthrough is given by:

$$N_p = \frac{0.6 \times 10^7}{1.25} = 4.8 \times 10^6 \text{ m}^3 \text{ (oil at stock tank conditions)}$$

Given that the injection rate has been constant, the time elapsed between the start of injection and the arrival of the front at the production wells is given by:

$$t_B = \frac{0.6 \times 10^7}{5\,000} = 1\,200 \text{ days}$$

Now let us calculate the performance of the system for $V_I = 1.5\ V_D$. From Fig. 15.22 we find that $E_s = 0.9$, thus:

$$N_p = \frac{0.9 \times 10^7}{1.25} = 7.2 \times 10^6 \text{ m}^3$$

and

$$t = \frac{10^7 \times 1.5}{5 \times 10^3} = 3\,000 \text{ days}$$

The results obtained show clearly the decline in oil production after breakthrough. Whereas at breakthrough 4.8 x 10^6 m^3 have been produced in 1 200 days, only a further 2.4 x 10^6 m^3 are produced in the following 1 800 days.

It is clear that if the fluid is injected at a constant rate, as in the preceeding example, the injection pressure will change throughout the displacement. If the mobility of the displacing fluid is greater than that of the displaced fluid, the pressure gradient in the swept area will be lower than that in the unswept area. The overall pressure gradient (from injector to producer) will thus vary with the area swept. To quantify this variation it is convenient to introduce the conductance ratio γ:

$$\gamma = (q/\Delta p)/(q/\Delta p)_i$$

where

the subscript i refers to the initial conditions,
q is the steady-state reservoir flowrate and,
Δp is the pressure difference between injector and producer.

Figures 15.23 and 15.24 show the relationship between γ and M for various values of E_s and V_I/V_D respectively, as obtained by Caudle and Witte for the five-spot pattern.

The following example illustrates the use of Fig. 15.23 and 15.24.

Example 1 (continued).

Initial Δp = 69 bar.
Initial reservoir flowrate = 31.8 m^3/day.

Two solutions are possible:

(a) Constant rate, variable injection pressure.
(b) Variable rate, constant injection pressure.

1) Constant rate:

With areal sweep efficiency at breakthrough of 0.6 and $M = 2.08$, Fig. 15.23 gives $\gamma = 1.4$. Thus we have:

$$\Delta p = (q/\gamma)/(q/\Delta p)_i = \frac{69}{1.4} = 49.3 \text{ bar}$$

For a cumulative volume injected of 1.5 times the displaceable pore volume the pressure drop will be:

$$\Delta p = \frac{69}{1.8} = 38.3 \text{ bar}$$

2) Variable rate, constant injection pressure:

At breakthrough the injection rate will be:

$$q = \gamma \Delta p (q/\Delta p)_i = 1.4 \times 31.8 = 44.5 \text{ m}^3\text{/day}$$

For a cumulative volume injected of 1.5 times the displaceable pore volume, the injection rate would be:

$$q = 1.8 \times 31.8 = 57.2 \text{ m}^3\text{/day}$$

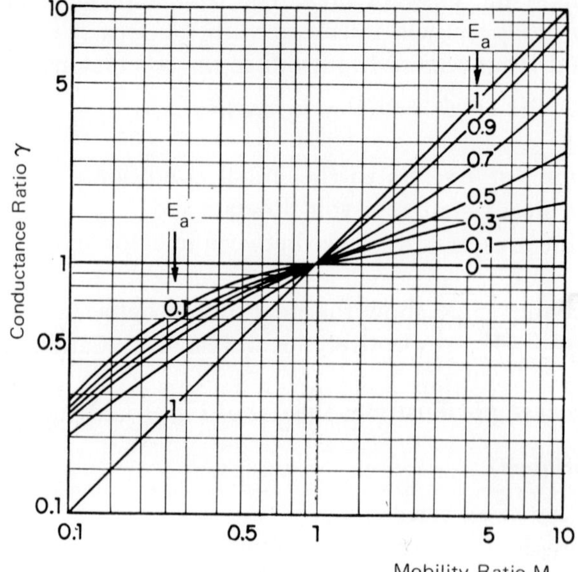

Fig. 15.23.

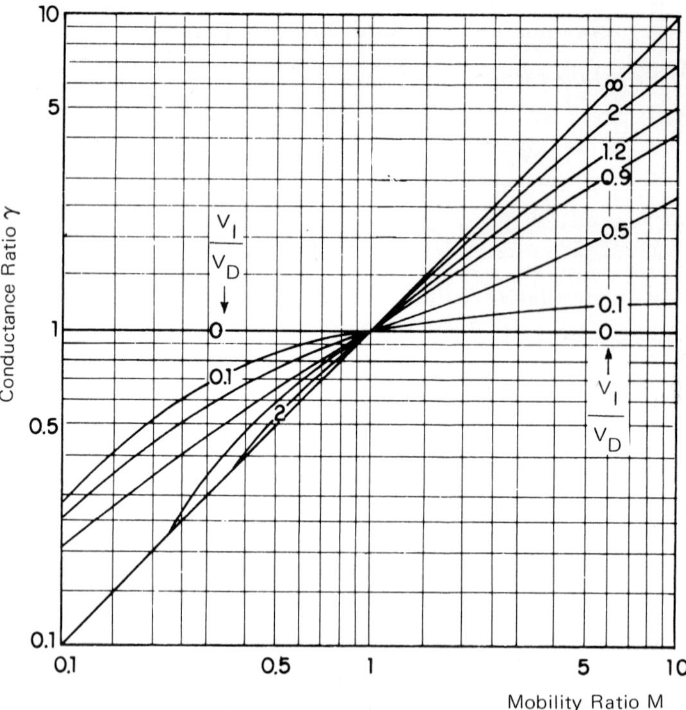

Fig. 15.24.

APPENDIX 1.1

The general theory of frontal displacement (Variable cross-section)

The problem to be studied is the displacement of oil by a non miscible fluid in the vicinity of an injection well completed in the oil zone of a homogeneous reservoir. As the production wells are some distance away the equipotentials are circular and centered on the injection well, and by symmetry they are also iso-saturation lines.

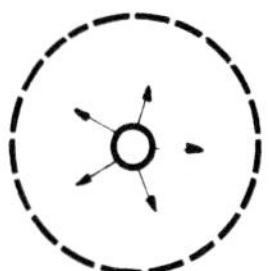

At a distance r from the wellbore the cross-sectional area offered to the fluids is $A = 2\pi rh$. Writing the equation of continuity between r and $r + \mathrm{d}r$ we have:

$$\frac{\partial}{\partial r}(f_1 Q_T) = -\frac{\partial}{\partial t}(A(r)\,\phi S_1)$$

Subscript 1 corresponds to the injected fluid. By a change of variables, using:

$$\xi = \frac{1}{A(r_0)} \int_{r_0}^{r} 2\pi hr \,\mathrm{d}r$$

the equation of continuity can be written:

$$\frac{\partial}{\partial \xi}(f_1 Q_T) = -\frac{\partial}{\partial t}(A(r_1)\,\phi S_1)$$

This equation is in the same form as that obtained for linear displacement, and the succeeding calculations are identical. Hence the equation for the velocity of an iso-saturation line is:

$$\left(\frac{\mathrm{d}\xi}{\mathrm{d}t}\right)_{S_1} = \frac{Q_T}{A(r_1)\,\phi}\left(\frac{\mathrm{d}f_1}{\mathrm{d}S_1}\right)_{S_1} \qquad (1)$$

After integration, this becomes:

$$\pi\phi[r^2(S_1, t) - r^2(S_1, 0)] = C_I(t)\left(\frac{df_1}{dS_1}\right) \tag{2}$$

Note that the left hand side of Eq. (2) represents the pore volume traversed by the iso-saturation S_1 at time t.

This result is completely general. As long as the displacement takes place in a porous medium whose geometry leads to the equipotentials and iso-saturations being superimposed, the change of variables $\xi = \frac{1}{A(0)} \int_0^x A(x)\,dx$ gives us Eq. (1). After integration the right hand side of Eq. (2) is again equal to the pore volume traversed by the iso-saturation.

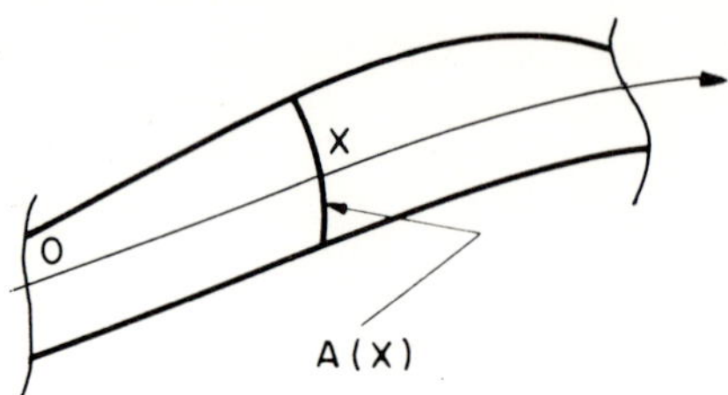

Marsal's General Theory

If the results of Buckley and Leverett are expressed in their most general form, a simpler theory of uni-dimensional displacement can be obtained. Ignoring the effects of gravity, Buckley-Leverett theory involves relative permeability only; all that is required is to include in the equation of continuity the condition that the fractional flow f_1 is a unique function of the saturation S_1. The Buckley-Leverett calculations then no longer require the concept of relative permeability.

D. Marsal noted that saturation is an intermediate parameter of relatively little practical use; in the study of fluid displacements the requirement is for forecasts of the relative production of oil and injected fluid, not of the fluid saturation at the wellbore.

Since f_1 is a unique function of saturation, Eq. (2) states that a given zone of fractional flow f_1 traverses a pore volume proportional to the cumulative volume injected. The results of non-miscible fluid displacements in a uni-dimensional porous medium can be perfectly represented by an experimental function of the following form:

$$\frac{1}{C_I(t)/V_p} = g(f_1) \tag{3}$$

where, for a linear displacement, ($V_p = \phi A x$):

$$x(S_1, t) - x(S_1, 0) = \frac{C_I(t)}{A\phi} g(f_1) \tag{4}$$

This is the simplest expression of the theory of linear displacement. It can be used to estimate the performance of upstructure wells from the observed performance of the lowest wells. Knowing that a fractional flow f_1 was observed at $x = L$ at time t after a cumulative injection $C_I(t)$, the same fractional flow f_1 will be observed at $x = L'$ at time t' when the cumulative injection C' downstream from L is such that:

$$\frac{C_I(t)}{V_p} = \frac{C_I'(t') - C_I'(t)}{V_p'} \tag{5}$$

where V_p is the pore volume between $x = 0$ and $x = L$ and V_p' that between $x = L$ and $x = L'$. Taking account of the production at $x = L$, Eq. (5) can be written:

$$\frac{C_I(t)}{V_p} = \frac{[C_I(t') - C_1(t') - C_2(t')] - [C_I(t) - C_1(t) - C_2(t)]}{V_p'} \tag{6}$$

where subscript 2 refers to oil production.

However, this simple description of a displacement does not include the prediction of the variation of pressure losses, which is of great importance in the study of fluid displacement in media of complex geometry.

REFERENCES

1 HOUPEURT, A., *Estimation des réserves par drainage naturel.* Cours de production, Tome V, Editions Technip, Paris, 1968.

2 DIETZ, D.N., *A theoretical approach to the problem of encroaching and by-passing edge water.* Konibel. Ned. Akad. Wetenschap, 1953, Proc. B 56.

3 LEFUR, B. and SOURIEAU, P., "Etude de l'écoulement diphasique dans une couche inclinée et dans un modèle rectangulaire de milieu poreux". Colloque ARTFP, *Rev. Inst. Franc. du Pétrole,* vol. XVIII, 6, 1963.

4 CHAUMET, P., COLONNA, J. and CROISSANT, R., "Stockage souterrain de gaz naturel en aquifère : dynamique de l'interface eau-gaz". *Rev. Inst. Franc. du Pétrole,* vol. XXI, 9, 1966.

5 SONIER, F., "Etude numérique du déplacement par perméabilités relatives des fluides non miscibles". Rapport IFP réf. 13 072, mars 1966.

6 CHAUMET, P. and SONIER, F. "Calculs numériques approchés des déplacements bidimensionnels". Communication n° 10 au séminaire ARTFP, Rueil, juin 1967.

7 PIRSON, S.J., *Oil reservoir engineering.* 2nd Ed., Mc Graw-Hill Book Co., Inc. New York, 1958.

8 MARLE, C., *Les écoulements polyphasiques en milieu poreux.* Cours de production, Tome IV. Editions Technip, Paris, 1972.

9 MOORE, T.F. and SLOBOD, R.L., *The effect of viscosity and capillarity on the displacement of oil by water.* Prod. Monthly, August 1965.

10 MUSKAT, M., *Physical principles of oil production.* Mc Graw-Hill Book Co., Inc., New York, 1949.

11 LANDRUM, B.L. and CRAWFORD, P.B., "Effect of directional permeability on sweep efficiency and production capacity". *JPT,* Nov. 1960.

12 ARNOLD, M.D., GONZALEZ, H.J. and CRAWFORD, P.B., "Estimation of reservoir anisotropy from production data". *JPT,* August 1962.

13 CHAUMET, P., Cours de récupération secondaire IAP. *Compléments de mécanique du balayage* (ronéoté), réf. 245, juin 1971.

14 CAUDLE, B.H. and WITTE, M.D., "Production potential changes during sweep-out in a five-spot system". *JPT,* Dec. 1959, p. 63.

2

water injection

21. INTRODUCTION

Water injection, the oldest assisted recovery method, remains the most common (80 % of the oil produced by enhanced recovery in the United States in 1970 was produced by water injection (Ref. 1)). Oil recovery is increased by an improvement in sweep or displacement efficiency.

In addition to the enhanced recovery objective, water injection may also be used in order to:

(1) Maintain the reservoir pressure when the expansion of the aquifer or gas-cap is insufficient for the purpose. In this instance the process should be regarded as one of pressure maintenance rather than of enhanced recovery. This is the case, for example, in the Parentis field in France and the Zelten field in Libya, in which water injection was not intended to improve the recovery but to increase the production rates. Thousands of other fields throughout the world employ water injection for this purpose.

(2) Dispose of the brine produced with the oil if surface discharge is not possible (e.g. into lakes or fresh water sources).

The water may be injected into an underlying or neighbouring aquifer.

22. THE SELECTION OF WATER INJECTION AS AN ENHANCED RECOVERY METHOD

When the natural reservoir energy is judged to be insufficient, the choice of enhanced recovery method is made according to both technical and economic criteria.

If more than one method is technically feasible, an economic analysis is performed for each one, comparing the increased revenue (due to the increase in oil recovery) with the additional expenditure required, and the method yielding the highest profit is chosen.

22.1. Technical factors

The two most common injection fluids are water and gas and we shall examine the advantages and disadvantages of them both.

Gas injection falls into two categories:

(a) Miscible gas injection.

(b) Non-miscible gas injection.

If conditions are technically suitable for miscible gas injection it may be the best method, since the microscopic displacement efficiency E_d will be improved.

In the case of non-miscible gas injection the following points should be noted:

Water injection is to be preferred in all cases where there are no practical constraints, due to the more favourable mobility ratio obtained.

In reservoirs containing highly under-saturated oil, water injection is all the more suitable since the low gas-oil ratios would result in only small volumes of gas being available for gas injection. Moreover, a considerable volume of gas has to be injected before a free gas phase can be formed in the reservoir.

In reservoirs containing saturated oil, water is the preferred injection fluid as long as the permeability to water is sufficiently high. However, in reservoirs containing volatile oil (very high GOR) other methods such as miscible gas injection may yield a higher recovery.

In heterogeneous water-wet reservoirs water injection is more efficient than gas injection due to the spontaneous imbibition of water, which does not occur with gas injection.

22.2. Economic factors

In the following sections of this Chapter we shall see how to evaluate the improvement in recovery efficiency due to water injection. In this section we shall examine the cost of an injection project, the various elements to be considered being:

(a) The cost of studies and laboratory work.

(b) The cost of drilling additional wells.

(c) The cost of converting producers into injectors.

(d) The capital and operating costs of the surface equipment: pumps, lines, tanks, filters, etc.

The capital costs of water injection systems are generally lower than those of gas injection systems.

The cost per cubic metre of injected water increases with the injection pressure, and for a given production rate a compromise must be found between using a few wells at high pressure or several wells at low pressure. Normally injection pressures are held below the formation fracture pressure, which averages 0.2 bar/metre.

The technical and economic results of studies comparing water injection with gas injection for the Gassi-Touil field are given in Appendix 2.1.

Another important factor in water injection is that there is a time lag between the start of injection and the increase in production (the oil rate increasing due to an increase in reservoir pressure or a decrease in the producing gas-oil ratio). The time-sensitive economic criteria (pay-out time, present worth, rate of return) are thus less favourable for waterflooding than they are for direct and immediate improvements to the productivity of production wells (gas-lift, paraffin cutting, workover, etc.). This is especially true for low permeability reservoirs, in which it may be more than a year after the start of injection that an appreciable increase in oil production is seen.

23. DISPLACEMENT MECHANICS

23.1. Homogeneous reservoirs

In this section we shall be assuming that the reservoir consists of a single homogeneous bed in which fluids move horizontally. This is a rather theoretical reservoir, but its study is however of value.

We shall assume that throughout the reservoir the saturation is constant.

Let us assume that at the start of a certain water injection project the reservoir has already been produced by natural depletion during the first phase of primary production. As is often the case, the current reservoir pressure is lower than the saturation pressure of the original oil in place. There will thus be a free gas phase present which, according to our hypothesis, will be assumed uniform throughout the reservoir.

It is possible that if the primary production phase was continued too long the oil will no longer be producible, since the gas-oil ratio will have reached too high a value at the production wells. At this stage only the gas will be mobile.

The commencement of water injection is accompanied by a pressure increase throughout the reservoir. The increase is greatest around the injection wells and declines towards the producers.

Under the influence of the increasing pressure the free gas tends to redissolve into the oil, with which it is in contact over a large surface area.

Consequently the oil production does not immediately increase following the start of water injection. There is initially a "fill-up" period during which a volume of water approximately equal to the volume of free gas initially present in the reservoir is injected. During the fill-up period a large proportion of the gas will be redissolved, the remainder being produced at the production wells. The fill-up can be represented by an oil front travelling ahead of and much faster than the water front; behind the oil front the gas saturation is at its residual value; the arrival of the oil front at the production wells marks the end of the fill-up period (Fig. 23.11). This phenomenon is not purely theoretical, it is generally experienced in fields produced by water injection, and the identity between the volume of water injected and the initial free gas volume has often been noted.

This volume may be estimated by a material balance calculation for the primary production phase, using the methods described in Volume V.

During the initial phase of injection, a cylindrical water front forms around each injection well, its radius increasing as injection continues.

Behind the water front the oil saturation is progressively reduced as more and more oil particles are caught in the moving stream of water, until finally the residual oil saturation is attained.

Laboratory experiments have shown that even if the initial interstitial water saturation is too low for water ahead of the front to move, a bank of interstitial water is nevertheless built up and pushed ahead of the injected water, and immediately behind the oil bank.

When fill-up has been achieved the frontal advance continues but the oil production rate increases, eventually becoming practically equal to the rate of water injection (in terms of reservoir volumes).

If the initial formation water saturation was lower than that required for it to flow, oil production during this second phase of injection will be water-free.

The start of significant water production signals the breakthrough of the front at the well concerned.

Water production increases rapidly at the expense of oil production; the gradual recovery of oil behind the front is only obtained by the circulation of considerable volumes of water. During this final phase of injection after breakthrough the area swept will also be increasing, and this may provide sufficient oil production to justify the continuation of injection. The project will come to an end once the operating costs exceed the income from the oil produced.

Economic criteria vary from place to place, but quite often the production wells are abandoned when the producing watercut reaches 95-98%.

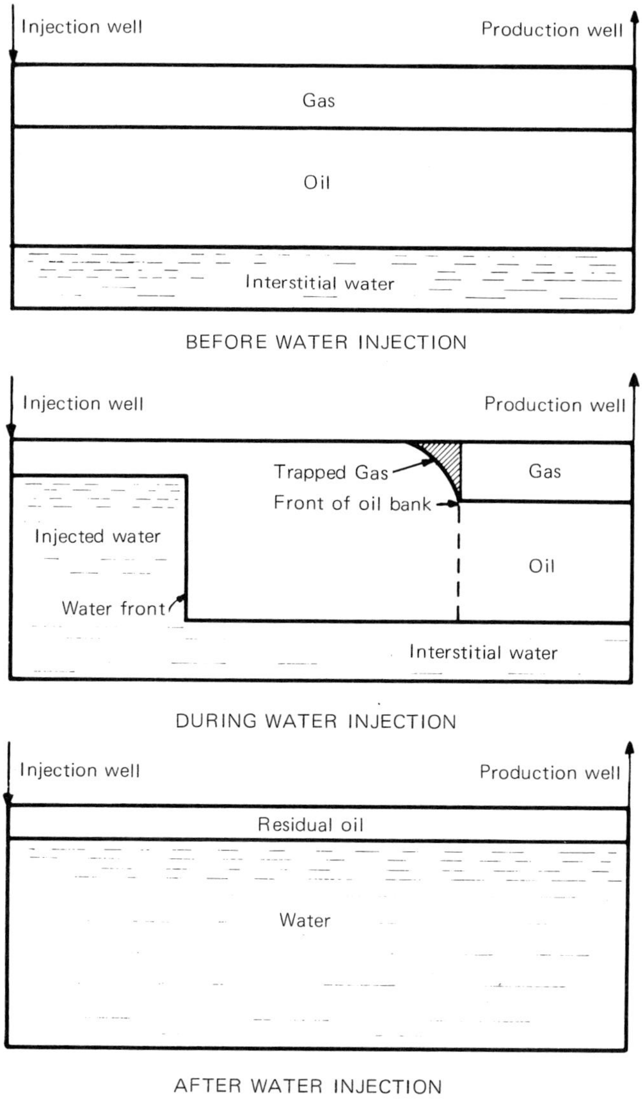

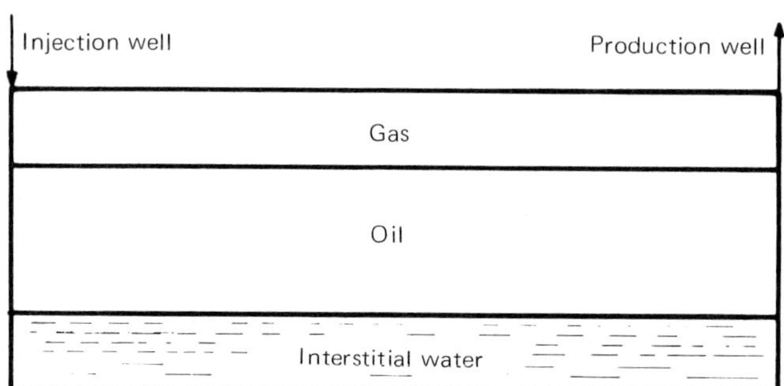

BEFORE WATER INJECTION

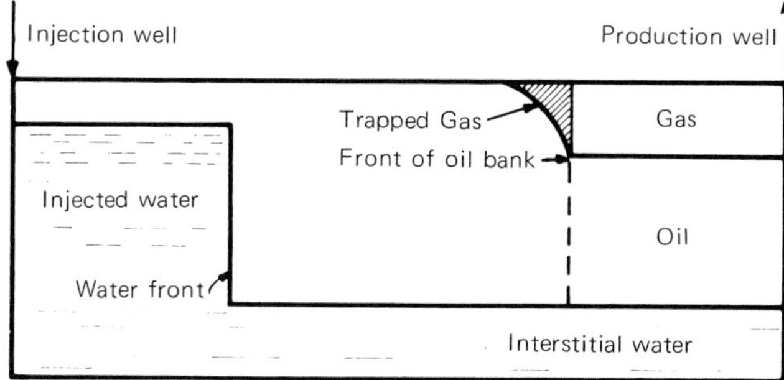

DURING WATER INJECTION

Injection well
Production well
Residual oil
Water

AFTER WATER INJECTION

$(S_{oi} - S_{or})\ V_p$ (of the zone swept by water)

$= S_{gi}\ V_p$ (of the oil bank)

where i signifies conditions at start of water injection

Fig. 23.11.

23.2. Heterogeneous reservoirs

Heterogeneous reservoirs may be divided into three principal types:

(a) Fissured reservoirs, in which one or more fracture systems divide the formation into more or less regular blocks and provide highly conductive fluid paths.

(b) Layered reservoirs, in which there are several parallel beds whose extent is usually great compared to their thickness, and which may or may not be in communication.

(c) Reservoirs with random heterogeneities, in which two or more types of porosity are distributed randomly.

For all of these types of reservoir (except a layered reservoir without communication) imbibition becomes significant the more the rock is preferentially water-wet. This phenomenon, resulting from the presence of capillary pressure gradients due to the partial invasion of the rock by the injected water, is usually augmented by gravity forces. Water tends to invade the less permeable zones and displace oil from them. Since the rate of imbibition is practically independent of the rate of water injection, imbibition becomes more significant when it has more time to take place, thus when the rate of displacement is slow.

Typical plots of oil recovery by water injection in heterogeneous reservoirs are shown in Fig. 23.21.

A. Fissured reservoirs

For fissured reservoirs it can often be assumed that the fluids are completely segregated by gravity in the fractures. During water injection there will thus be a steadily rising oil/water interface in the fractures which remains reasonably flat and horizontal (Fig. 23.22). In this case it is possible to describe the problem mathematically with a fair degree of accuracy and thus, as long as the size of the blocks can be determined, reliable performance predictions can be made. In some cases it is possible to estimate the block size by history matching.

B. Layered reservoirs

For layered reservoirs there are two cases to be considered, depending on whether or not there is interlayer communication.

1. Reservoirs without interlayer communication

In this case the description given in Section 23.1 remains valid for each individual layer. The water front in each layer advances at a velocity dependent on the total injection rate and the properties of the layer.

This explains why, if free gas is present at the start of injection, fill-up of the

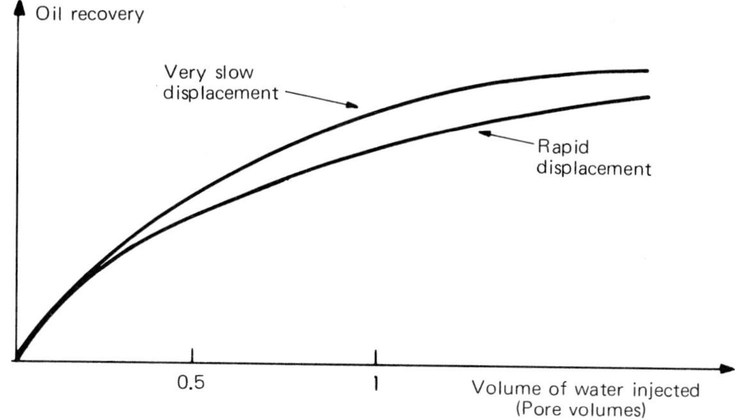

Fig. 23.21.

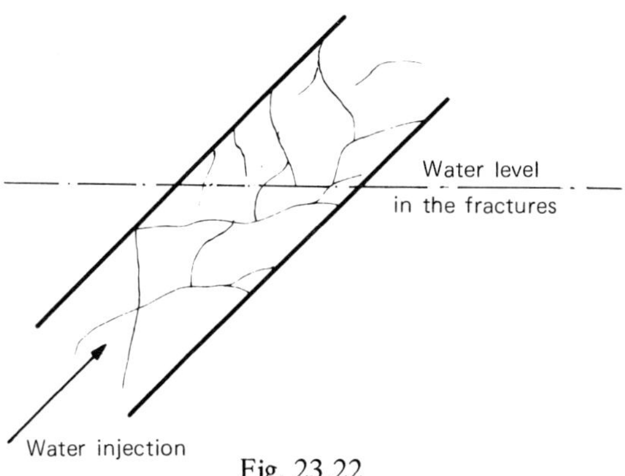

Fig. 23.22

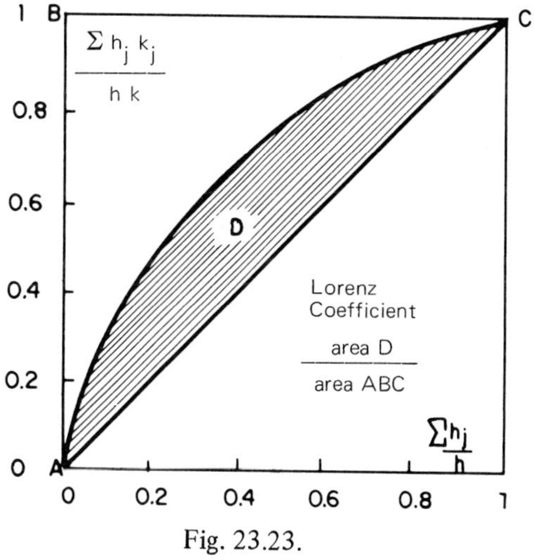

Fig. 23.23.

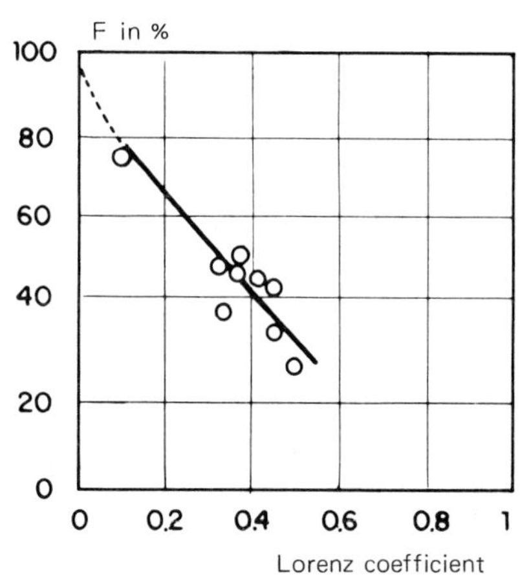

Fig. 23.24.

most permeable layers occurs first and oil production increases when a volume of water less than the free gas volume has been injected ($F\%$ of the latter volume).

The work of Schauer, Schmalz-Ralme and Lorenz enables us to estimate the fraction F of the free gas volume that it is necessary to replace by water before a change in oil production will be observed.

Figure 23.24 gives F as a function of the Lorenz coefficient which itself is calculated from the permeability distribution as shown in Figure 23.23. The Lorenz coefficient is a measure of the heterogeneity of a reservoir. For a homogeneous formation the Lorenz coefficient is zero.

2. Reservoirs with interlayer communication

The problem of the displacement of oil by water in a formation consisting of two parallel communicating beds (Fig. 23.25) has been studied both experimentally and theoretically (*cf* Volume IV of this Production Course).

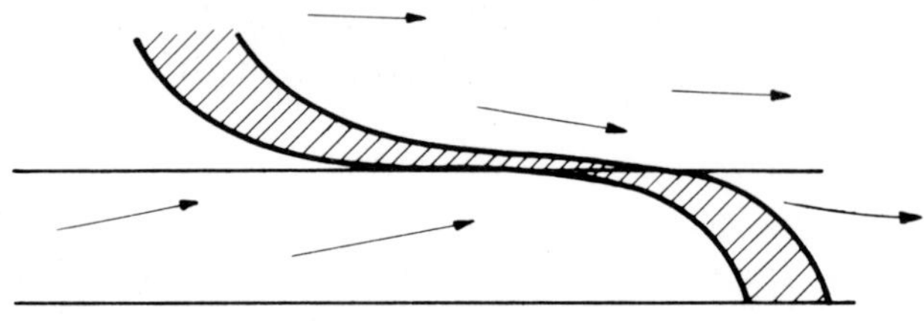

Fig. 23.25.

If we take account of capillary effects, we shall have transition zones instead of distinct fronts. It has been observed that the transition zone initially moves fastest in the most permeable layer, but that after a certain distance has been covered the separation between the transition zones remains roughly constant. The magnitude of the separation is greatest at high injection rates, which is to be expected, since the effects of imbibition are greatest when the injection rate is low. Correlations are available which enable the separation between the transition zones to be calculated, from which the quantity of oil remaining in the least permeable layer at breakthrough in the most permeable layer can be estimated.

C. Reservoirs with random heterogeneities

These reservoirs are those in which two or more types of porosity are distributed in a somewhat complicated fashion which varies considerably from one reservoir to another (Fig. 23.26).

In this type of reservoir, imbibition also takes place. As for the previous cases, the relationship between oil recovery and water injected has the form shown in Fig. 23.21. It is possible to take account of imbibition theoretically, by ascribing to the reservoir average relative permeabilities which are functions of the flowrate. But, in practice, we very rarely have sufficiently precise information about the distribution of the heterogeneities to be able to take account of them effectively.

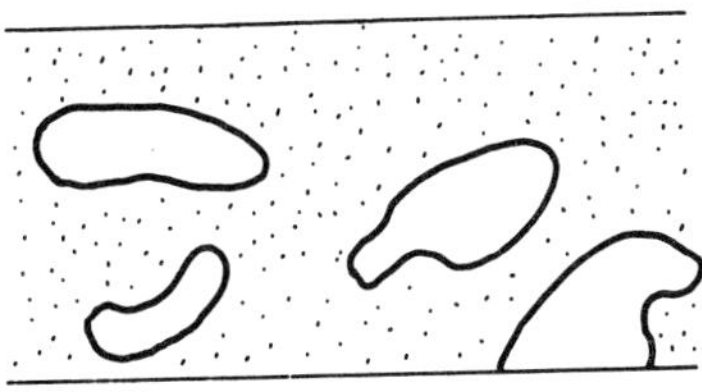

Fig. 23.26.

24. WATER INJECTION IN REGULARLY DEVELOPED HOMOGENEOUS RESERVOIRS

Consider a homogeneous reservoir in which during the initial natural depletion stage of development the pressure has fallen far below the bubble-point. Water injection has just started, being a regular pattern flood at constant pressure.

Around each injection well of radius r_w an oil bank forms, defined by two cylindrical surfaces of radii r_0 and r_e (Fig. 24.1). During fill-up, oil offtake is very limited. The pressure gradients between injection and production wells are therefore small and the fronts maintain their cylindrical shape for a considerable time.

After a certain time the oil banks of two neighbouring injection wells join together. This is known as interference (Fig. 24.2).

The injected volumes at which interference occurs (see notation on Fig. 14.21 to 14.23) are:

Direct line drive:

$$\overline{W_i} = \frac{1}{4}\left(\frac{a}{r_w}\right)^2 \qquad a \leqslant 2d$$

$$\overline{W_i} = \left(\frac{d}{r_w}\right)^2 \qquad a \geqslant 2d$$

Five-spot:

$$\overline{W_i} = \frac{1}{2}\left(\frac{L}{r_w}\right)^2$$

Seven-spot:

$$\overline{W_i} = \frac{1}{4}\left(\frac{L}{r_w}\right)^2$$

where

$\overline{W_i}$ is the dimensionless injected volume:

$$\overline{W_i} = \frac{w_i}{\pi h \phi_w r_w^2}\left(\frac{\phi_w}{\phi_o}\right)^M$$

and

$$\phi_w = \phi(1 - S_{wi} - S_{or})$$
$$\phi_o = \phi(1 - S_{wi} - S_{oi})$$

From the time that interference initially occurs, fill-up continues until there is no longer any free gas in the reservoir. Subsequently, the sweep continues in the liquid phase until the injected water breaks through at the production wells.

There are thus three successive stages:

(a) Fill-up before interference.
(b) Fill-up after interference.
(c) Injection after fill-up.

The change in injection rate with time has been calculated by Muskat, who obtained the following results for a constant injection pressure (Fig. 24.3):

1. Fill-up before interference

The injection rate decreases very rapidly at the start and very slowly at the end of this period, according to the relationship:

$$1 + \exp^{(1/\overline{Q})}\left(\frac{1}{\overline{Q}} - 1\right) = \bar{t}$$

where

$$\overline{Q} = \frac{Q\,\mu_w}{4\pi\, k_w\, h\, \Delta P}$$

and

$$\bar{t} = \frac{4k_w\, t\, \Delta P}{\mu_w\, \phi_w\, r_w^2}\left(\frac{\phi_w}{\phi_o}\right)^M$$

2. Fill-up after interference

This period is too complex to be treated analytically. For practical purposes a linear decline in injection rate can be assumed.

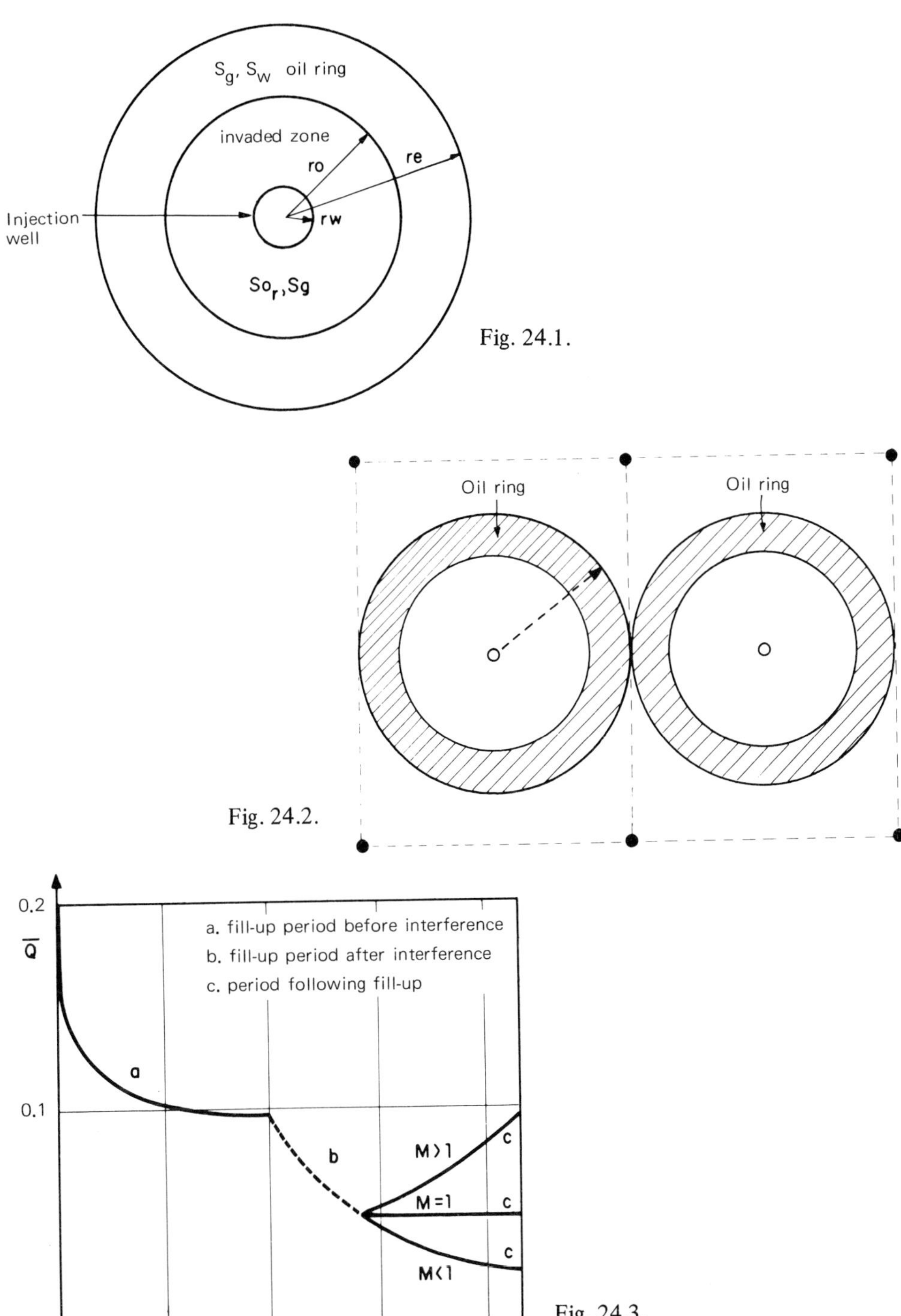

Fig. 24.1.

Fig. 24.2.

Fig. 24.3.

3. Injection after fill-up

After fill-up the change in injection rate depends on the mobility ratio M between the injected water and the displaced oil. If M is unity the rate remains constant, if $M > 1$ it increases and if $M < 1$ it decreases, having the same form as γ in Fig. 12.24.

25. WATER INJECTION PERFORMANCE CALCULATIONS

The results required are estimates of final oil recovery and the oil and water production rates.

The amount of oil recoverable by water injection can be calculated by the equation:

$$N_p = \frac{V_p (S_{oi} - S_{or}) E_s E_i}{B_o}$$

where

- V_p is the swept pore volume. This may be rather smaller than the pore volume which contributed to the natural recovery phase. This is especially true for reservoirs with lenticular zones,
- S_{oi} is the initial oil saturation at the start of water injection,
- S_{or} is the residual oil saturation at the end of water injection,
- E_s is the superficial or areal sweep efficiency (see Chapter 1),
- E_i is the invasion efficiency or vertical sweep efficiency.

In general, recovery by water injection is of the order of 30 to 50 % of the initial oil in place for reservoirs consisting only of matrix porosity and containing under-saturated oil. Recovery can be much lower than this in fissured reservoirs, reservoirs already depleted by dissolved gas drive or those in which there are very few wells.

Forecasts of oil and water production rates may be made manually or by the use of analogical or mathematical models.

25.1. Hand calculations. Analytical methods.

The following calculation methods apply to linear systems and result in a value for the fractional recovery R. To extend this result to other geometrical configurations, R should be multiplied by E_v (see Chapter 1).

We shall discuss in turn the Buckley-Leverett and viscous fingering methods for homogeneous reservoirs, and the Stiles, Dykstra-Parsons and Johnson methods for layered reservoirs. Finally various other existing methods will be briefly mentioned.

25.11. Homogeneous reservoirs

A. Buckley-Leverett theory

This theory is correct when $M < 1$ and may generally be used up to $M = 10$.

The method of calculating the time to breakthrough and the recovery for a given saturation at the production wells was shown in Section 12.1.

Two separate periods must be considered in the calculation of oil rate and producing water/oil ratio (WOR): the periods before and after breakthrough (Fig. 25.111).

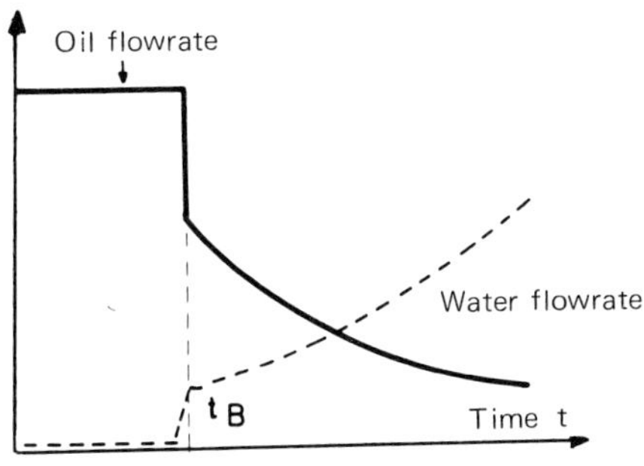

Fig. 25.111.

1. Pre-breakthrough period

Oil production is constant, $Q_o = Q_T$, and the WOR is zero if the water saturation at the start of water injection is equal to the irreducible or minimum water saturation S_{wm}. If $S_{wi} > S_{wm}$ however, there will be a certain WOR from the outset.

2. Post-breakthrough period

At a certain time t_1, the water saturation around a production well is the unknown y, given by the solution of the implicit equation:

$$\left(\frac{\mathrm{d}f_w}{\mathrm{d}S_w}\right)_{S_w = y} = \frac{L}{\dfrac{Q_T}{A\phi}\, t_1} = \frac{V_P}{Q_T t_1}$$

Thus y may be determined graphically.

Having found the saturation S_w, f_w may be calculated and thus the WOR determined. Note that f_w is in terms of bottom-hole conditions whilst the water/oil ratio (WOR) is calculated at surface conditions, thus:

$$\mathrm{WOR} = \frac{f_w B_o}{(1 - f_w) B_w}$$

Produced water can be re-injected (thus reducing the water supply required) and very often this is obligatory. Produced waters are brines of various salinities contaminated with traces of emulsified hydrocarbon, and their discharge at surface may be prohibited by law.

Note

The necessary condition for the formation of an oil bank is that the sum of the initial water and gas saturations lies between S_{wm} and the saturation S_{wF} (see Fig. 25.112). The initial water and gas saturations are summed because of the logical assumption that until fill-up is complete the injected water takes the place of the free gas.

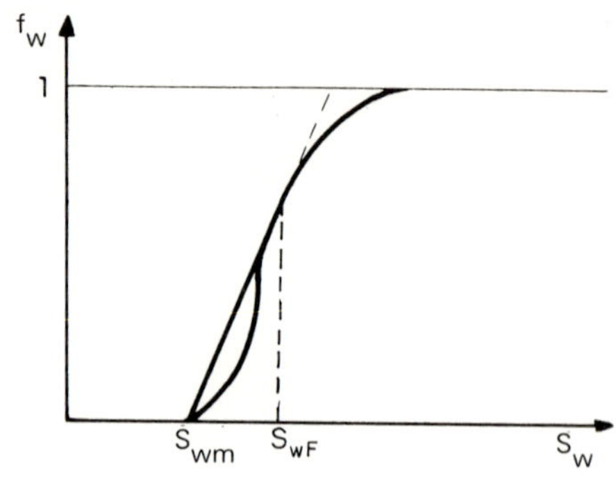

Fig. 25.112.

B. *Viscous fingering theory*

When the mobility ratio is "unfavourable" (greater than 10) the Buckley-Leverett method no longer applies and the viscous fingering method may be used instead. Van Meurs and Van Der Poël developed the theory, which is based on the following assumptions:

(1) The flow may be split into three parts, zone 1 in which only oil is flowing, zone 2 in which only water is flowing and zone 3 in which both water and oil are immobile (Fig. 25.113).

(2) Zones 1, 2 and 3 may be characterized by their oil and water saturations:

. Zone 1: $S_o - S_{or} = 1 - S_{or} - S_w$
. Zone 2: $S_w - S_{wm}$
. Zone 3: $S_{wm} + S_{or}$

(3) The system is at steady-state conditions.
(4) The formation is horizontal and homogeneous.
(5) The fractions of oil and water flowing in any given part of the system are space and time independent.

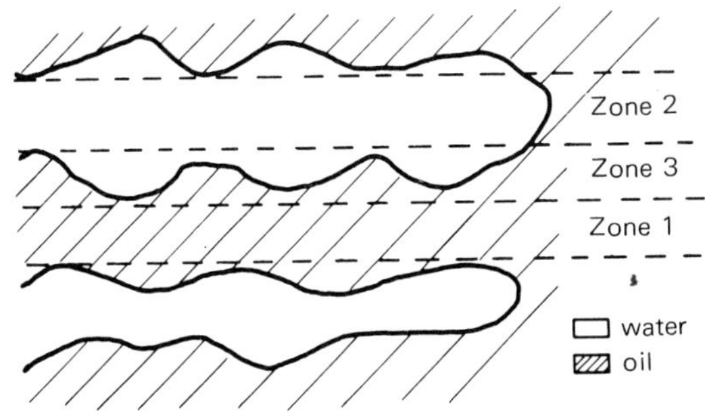

Fig. 25.113.

In zone 2, let u_w be the filtration velocity of the water:

$$u_w = -\frac{k}{\mu_w}\frac{\partial P_w}{\partial x} \qquad \text{(Eq. 25.111)}$$

Only water is flowing, thus:

$$q_w = u_w(S_w - S_{wm}) \qquad \text{(Eq. 25.112)}$$

Similarly in zone 1, in which only oil is flowing:

$$u_o = -\frac{k}{\mu_o}\frac{\partial P_o}{\partial x} \qquad \text{(Eq. 25.113)}$$

and

$$q_o = u_o(1 - S_{or} - S_w) \qquad \text{(Eq. 25.114)}$$

If we assume that $P_0 - P_w$ = constant,

$$\frac{\partial P_o}{\partial x} = \frac{\partial P_w}{\partial x}$$

we then have:

$$u_w = u_o M \qquad \text{(Eq. 25.115)}$$

and

$$q_w = u_o M (S_w - S_{wm})$$
$$q_o = u_o (1 - S_{or} - S_{wm} + S_{wm} + S_w)$$
$$q_T = u_o (M - 1)(S_w - S_{wm}) + u_o B$$
$$f_w = \frac{q_w}{q_T} = \frac{M (S_w - S_{wm})}{(M - 1)(S_w - S_{wm}) + B}$$

where B is the constant $(1 - S_{or} - S_{wm})$.

After some straightforward calculation which we shall not develop here, the authors obtain the following principal results:

Oil recovery at breakthrough is given by:

$$N_{pB} = \frac{[B + \sqrt{(M - 1) S_{wm} B}]^2}{MB}$$

The length over which all the displaceable oil has been swept is given by:

$$X_1 = \frac{W_i}{BM}$$

where

W_i represents the cumulative volume of water injected, expressed in pore volumes.

The fractional flow of water at breakthrough is given by:

$$f_{wB} = \frac{M}{M - 1}\left(1 - \sqrt{\frac{B}{MW_i}}\right)$$

The oil recovery for any given limiting water cut f_l is given by:

$$N_{pl} = S_{wm} + B \frac{[M - f_1^2 (M - 1)]}{[M - f_l (M - 1)]^2}$$

25.12. Layered reservoirs

A. The Stiles method

This method only applies when M is close to unity.

In his calculations for a layered reservoir, Stiles makes the following assuptions:

(a) The formation is made up of a number of continuous beds of constant thickness.

(b) There is no fluid segregation within a bed or communication between beds.

(c) The displacement is "piston-like".

(d) The system is linear, with the same relative permeability to oil ahead of the front and to water behind the front in each bed. Apart from the absolute permeability, the beds have identical petrophysical properties (porosity, saturations, etc.).

(e) The position of the front in each bed is directly proportional to the absolute permeability of the bed.

(f) The percentage of water in the production depends on the kh of the beds in which water has already broken through compared to the total kh of the system.

In view of these assumptions it is convenient to divide the reservoir into a number of layers of constant thickness h_i, h_r, etc., and of permeability k_i, k_r, etc., and to number them in order of decreasing permeability. The first layer in this model will therefore be the first to be completely swept and so on.

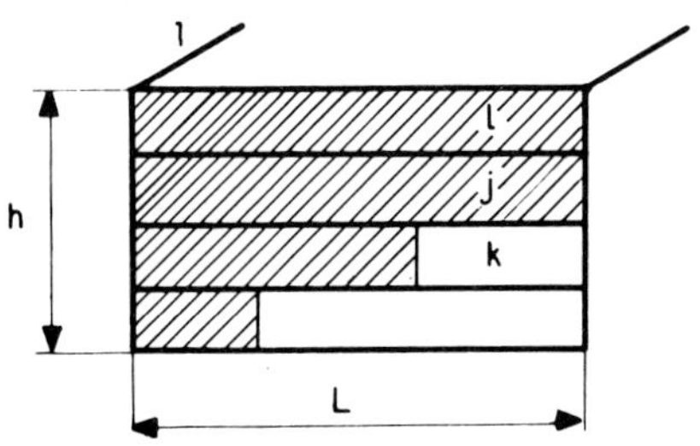

The total volume of displaceable oil is:

$$L l \phi h \, (1 - S_w - S_{or})$$

The volume displaceable from the ith layer is:

$$L l \phi \, h_i \, (1 - S_w - S_{or})$$

When the front breaks through in the jth layer, in a less permeable layer (thus one of a higher index) such as the kth, it will have travelled a distance $L \dfrac{k_k}{k_j}$. The volume of oil produced from the kth layer is thus:

$$L l \phi h_k \frac{k_k}{k_j} (1 - S_w - S_{or})$$

The total volume of oil produced at the time of breakthrough in the jth layer is:

$$L l \phi \left[1 - S_w - S_{or}\right] \times \left[\sum_1^j h_i + \sum_{j+1}^n \frac{k_i}{k_j} h_i\right]$$

Thus the recovery of displaceable oil is given by:

$$R = \frac{1}{h}\left[\sum_1^j h_i + \frac{1}{k_j}\sum_{j+1}^n k_i h_i\right]$$

This equation may also be written:

$$R = \frac{\sum_1^j h_i}{h} + \frac{(C_t - C_j)}{k_j h}$$

where

C_t is the total kh of the formation,

C_j is the cumulative kh up to the jth layer.

If we define f_w and f'_w as the fractional flow of water at bottom-hole and surface conditions respectively, it can easily be shown that at breakthrough in the jth layer:

$$f_w = \frac{MC_j}{MC_j + (C_t - C_j)} \quad \text{where } M = \frac{k_{rw}}{k_{ro}} \frac{\mu_o}{\mu_w}$$

$$f'_w = \frac{AC_j}{AC_j + (C_t - C_j)} \quad \text{where } A = \frac{k_{rw}}{k_{ro}} \frac{\mu_o}{\mu_w} \frac{B_o}{B_w}$$

An example of the application of the Stiles method

Let us consider a reservoir consisting of three layers containing 2 000 m^3 of displaceable oil at stock tank conditions. This volume is based on 3 150 m^3 of stock tank oil in place, initial oil saturation 0.6, residual oil saturation 0.2, areal sweep efficiency 0.95.

The oil formation volume factor is 1.073, initial water saturation 0.23 and mobility ratio $M = 1.32$.

The injection rate is held constant at 20 m^3/day.

Most of the calculations are shown in Table 25.121, but those for the fill-up period are shown below.

TABLE 25.121

(1) h_j Layer thickness (m)	(2) Σh_j Cumulative thickness from the top	(3) k_j (mD)	(4) $\Sigma k_j h_j$ or C_j	(5) $k_j(\Sigma h_j)$	(6) R	(7) N_p $(2000 \times R)$ Stock tank oil (m^3)	(8) f_w After breakthrough (at bottom hole conditions)	(9) Oil flow rate before breakthrough (stock tank) $18.6 \times (1 - (8)_{j-1})$ (m^3/day)	(10) $t = 19 + \Delta \frac{(7)}{(9)}$ (days)	(11) W_i (m^3)	(12) f'_w After breakthrough (at surface conditions)
1	1	310	310	310	0.602	1 204	0.622	18.60	84	1 680	0.638
1	2	187	497	374	0.778	1 556	0.914	7.03	134	2 680	0.918
1	3	63	560	189	1.000	2 000	1.000	1.60	411	8 220	1.000
Total 3											

Note: The recovery R may be written for this case as:

$$R = \frac{0.333}{k_j} (k_j(\Sigma h_j) + 560 - C_j)$$

For example, at breakthrough in the second layer:

$$R = \frac{0.333}{187} (374 + 560 - 497) = 0.778 \qquad f_w = \frac{1.32\, C_j}{0.32\, C_j + C_t} \qquad f'_w = \frac{1.416\, C_j}{0.416\, C_j + C_t}$$

The volume initially occupied by gas is:

$$\frac{2\,000 \times 1.073}{0.4 \times 0.95} \times (1 - 0.6 - 0.23) = 960 \text{ m}^3$$

Let us assume that the Lorenz coefficient for this reservoir has been calculated and from Fig. 23.24 a value of $F = 0.4$ has been obtained. There will thus be no visible reaction to water injection until a volume of water equal to 0.4 times the volume of free gas has been injected, being $960 \times 0.4 = 384 \text{ m}^3$. The fill up period will thus last about $384/20 = 19$ days.

B. *The Dykstra-Parsons and Johnson methods*

Stiles' calculations do not take into account the continual variation of the injectivity of each layer according to the advance of the water front. However, the Dykstra-Parsons method overcomes this problem, and is valid for a wide range of mobility ratios. It is based once again on the theory of "piston-like" displacement.

As for the Stiles method, let us consider the moment of breakthrough in the ith layer.

From Eq. 12.21 of Chapter 1 we have:

$$-\frac{t_i \Delta p}{\phi_D} = \frac{X^2}{2\lambda_{wi}}(1 + M)$$

Substituting this value of t_i in Eq. 12.22 applied to a lower permeability layer (the jth) we have:

$$\left(\frac{x}{X}\right)_j = \frac{-M + \sqrt{M^2 + \frac{k_j}{k_i}(1 - M^2)}}{1 - M}$$

The ratio $(x/X)_j$ is the recovery factor for layer j at time t_i.

For a system of n layers, the recovery at breakthrough in the mth layer is:

$$\frac{1}{n}\left[m + \frac{n - m}{M - 1} M - \frac{1}{M - 1} \sum_{m+1}^{n} \sqrt{M^2 + \frac{k_i}{k_m}(1 - M^2)}\right]$$

At the same time, the ratio of water flowrate to oil flowrate at bottom-hole conditions is:

$$\sum_1^m k_i \Big/ \sum_{m+1}^{n} \frac{k_i}{\sqrt{M^2 + \frac{k_i}{k_m}(1 - M^2)}}$$

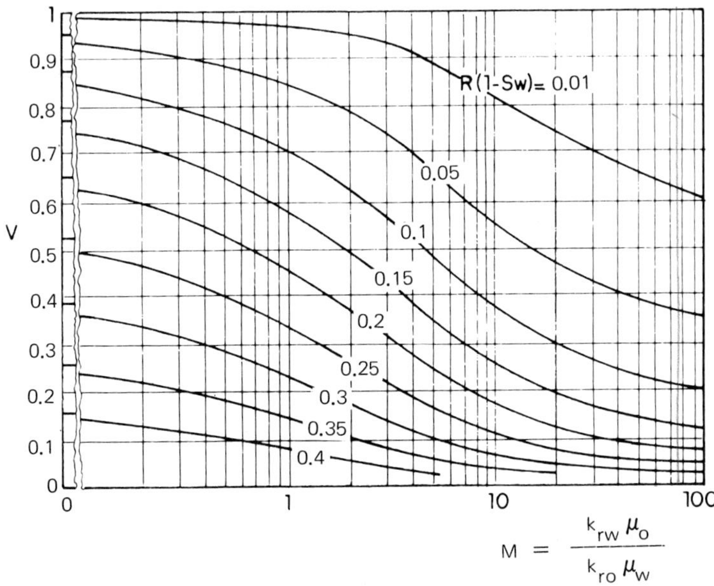

Fig. 25.121. Johnson's correlation for a producing WOR of 1.

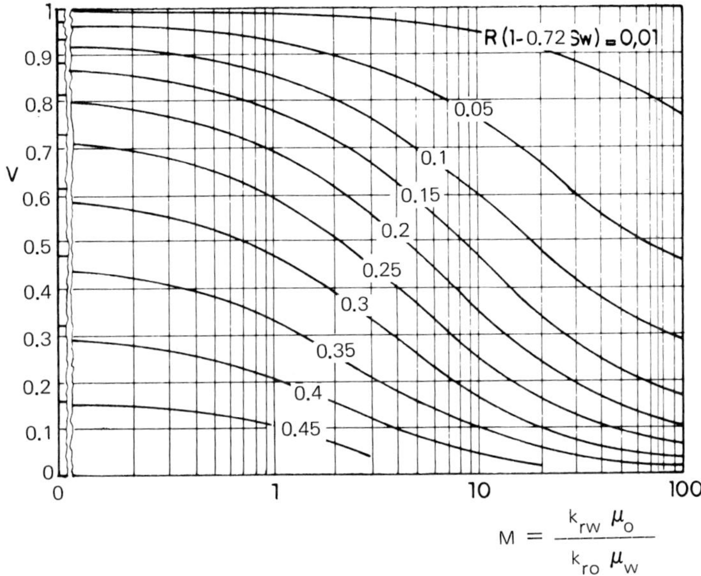

Fig. 25.122. Johnson's correlation for a producing WOR of 5.

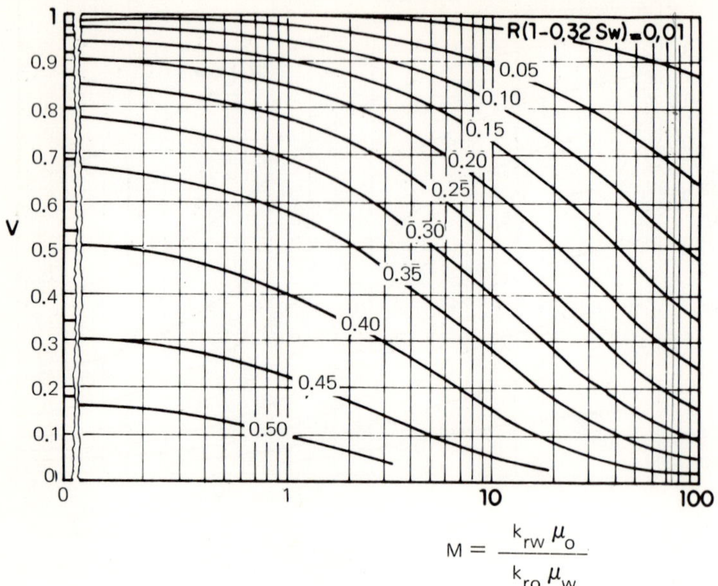

Fig. 25.123. Johnson's correlation for a producing WOR of 25.

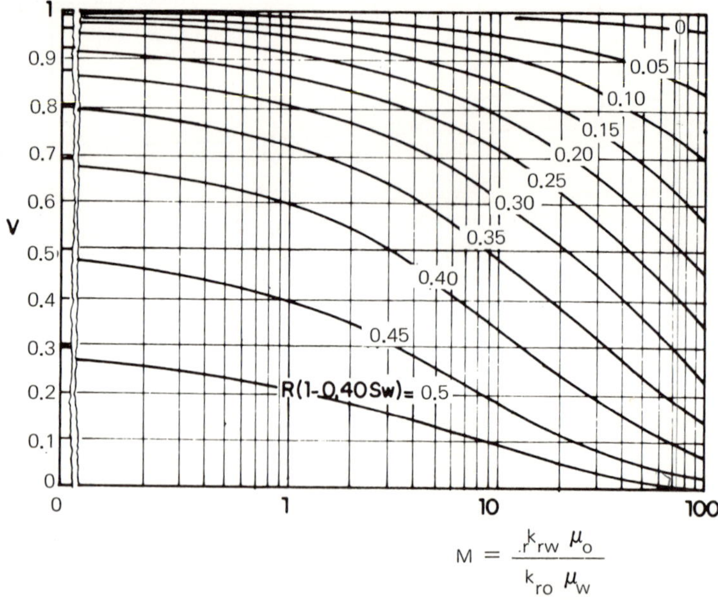

Fig. 25.124. Johnson's correlation for a producing WOR of 100.

Johnson has presented a simplified graphical approach to the Dykstra-Parsons method for the case of a formation with a log-normal or "Gaussian" permeability distribution, characterised by its variance. Figures 25.121 to 25.124 show the correlations obtained between the vertical variation of permeability V, the initial water saturation S_w, the mobility ratio M and the oil recovery R (fraction of initial oil in place), for various values of the producing water/oil ratio.

V is calculated from the statistical analysis of the permeability distribution by plotting the permeability values on log probability paper and choosing the best straight line through the points.

If $k_{84.1}$ is the permeability read from the line such that 84.1% of the permeability values are greater than $k_{84.1}$, V can be defined as follows:

$$V = \frac{k_{84.1} - k_{\text{median}}}{k_{\text{median}}}$$

Example of the use of Johnson's curves

Consider a reservoir with $V = 0.3$, $S_w = 0.30$ and $M = 2.6$ in which the primary recovery was 10 %. From the curves we have:

WOR	Reading from curve	Total recovery factor R	Recovery due to water injection
1	$R(1 - S_w) = 0.23$	0.33	0.23
5	$R(1 - 0.72S_w) = 0.33$	0.42	0.32
25	$R(1 - 0.52S_w) = 0.41$	0.49	0.39
100	$R(1 - 0.40S_w) = 0.46$	0.52	0.42

These results are plotted on Fig. 25.125. If the economic limit of production was a water/oil ratio of 25 an increase in recovery to 39% by water injection could be predicted.

The Johnson method is thus a fast way of obtaining an idea of the profitability of an injection project.

It gives reasonable results when the initial oil saturation is more than 45%.

25.13. Other methods

Amongst the many existing methods the following may be noted:

(a) Craig, Geffen and Morse method.
(b) Abernathy (modified Craig) method. This method allows the assumption

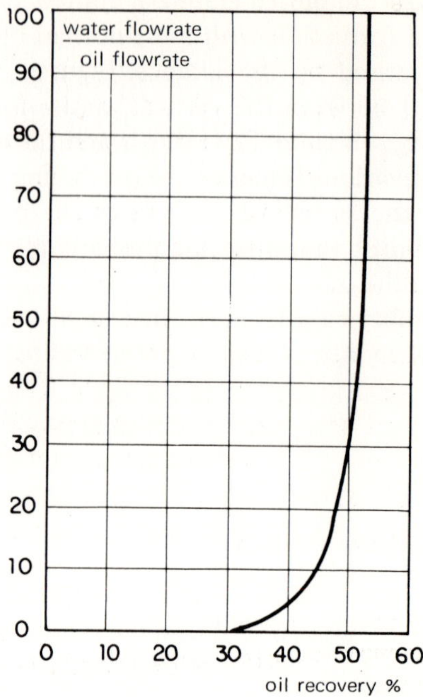

Fig. 25.125. An example of the use of Johnson's correlation.

of beds of non-uniform porosity, permeability and relative permeability. It is rather complex and is better handled by computer than manually.

(c) Carter Oil Company method.

25.2. Hand calculations. Empirical methods

If water injection is in progress in part of a reservoir or in neighbouring reservoirs, the performance data obtained may be used in the predictions for the so far unswept part of the reservoir. These empirical forecasts are made by the calculation of final recovery, the ratio $\frac{\text{production rate}}{\text{injection rate}}$ as a function of cumulative injection, and the average injection rate. To estimate these quantities for the zone to be studied, the principle of proportionality is used, relating the total oil recovery (primary and enhanced) to the oil initially in place, and the flowrates to the kh and applied pressure difference.

Finally, extrapolation of the oil rate or water injection rate in a reservoir is now at a sophisticated level, and is a valuable forecasting tool, as it is for all other production methods.

25.3. Mathematical models

Mathematical models have today almost totally replaced analogue models (based on the similarity of electrical laws and the laws of fluid flow in porous media), due to the very high speed of modern computers.

Existing mathematical models (2 and 3 dimensional, 2 and 3 phase) are well suited to the solution of the problems of recovery by water injection.

Great care must be taken in the preparation of data for entry into a computer-based reservoir model. It has already been noted that V_p may differ from the pore volume considered during primary depletion. It is also important to consider carefully the values used for the relative permeabilities.

If pressure behaviour is to be simulated, the use of a single phase program is recommended, being relatively cheap to run. The use of 3-dimensional programs is not recommended unless the layer properties are sufficiently well known.

The calculation of production and injection well capacities may be appended to the reservoir model program if they are not already included (productivity and injectivity indices, pump and flowline characteristics). The objectives of the model study (which may encompass more than the prediction of reservoir performance: additional drilling, conversion of production wells into injectors, etc.) must be very limited and very well defined, since the cost per run is generally high.

The successful use of these models requires careful consideration of the type of model and data to be used, and the writing and operation of mathematical reservoir models demands specialised training.

26. OPTIMUM INITIAL FREE GAS SATURATION FOR WATER INJECTION

Numerous researchers (Refs. 9 to 15) have demonstrated the beneficial effect of the presence of an initial free gas saturation on the final oil recovery by water injection. The effect reaches a maximum for a certain value of gas saturation, and the recovery then obtained is rather greater than it would be if the field was operated above the bubble-point.

The theory of this phenomenon is not well established. According to Cole (Ref. 16) the explanation lies in the fact that the interfacial tension between water and gas is higher than that between gas and oil. Since the reservoir fluids have a tendency to arrange themselves so as to minimise their energy, most of the gas particles surround the oil. Consequently, the residual oil saturation after waterflooding is reduced.

From laboratory results (Ref. 8) the following correlation has been derived giving the optimum gas saturation:

$$(S_g)_{opt} = \frac{0.376\, k^{0.634}\, B_o^{0.902}}{\left(\dfrac{S_o}{\mu_o}\right)^{0.352} \left(\dfrac{S_w}{\mu_w}\right)^{0.166} \phi^{1.152}} \qquad \text{(Eq. 26.1)}$$

In this equation k is in mD, μ in cP, S_o and S_w in fractions and ϕ in percent (for 20% porosity, $\phi = 20$).

This correlation is not explicit and must be used in conjunction with a material balance equation. When the gas saturation given by material balance and by the correlation are identical the gas saturation determined is the optimum value.

Example of the use of Eq. 26.1.

Consider a reservoir of permeability $k = 33$ mD, porosity $\phi = 25\%$, initial water saturation $S_{wc} = 0.3$ and of which the fluid properties below the saturation pressure are the following:

Pressure (bar)	S_o (%)	B_o (m^3/m^3)	μ_o (cP)	S_g (%)
175	70.0	1.333	0.600	0.0
160	62.8	1.287	0.625	7.2
140	56.8	1.250	0.650	13.2
122	52.7	1.221	0.700	17.3

If the correlation Eq. 26.1 is applied for a pressure of 160 bar, the calculated $(S_g)_{opt}$ is 12.2% which does not correspond to the actual saturation in the reservoir of 7.2%. The process is thus one of successive approximation.

For a pressure of 140 bar the calculated $(S_g)_{opt}$ is 12.2% which agrees with the reservoir gas saturation.

Thus to obtain the highest recovery by water injection in this reservoir, the natural depletion phase should be terminated at a pressure of 140 bar and water injection at constant pressure initiated.

27. PRACTICAL CONSIDERATIONS IN WATER INJECTION PROJECTS

In order to put water injection into operation in a reservoir the following points must be considered:

(a) Injection well completions.
(b) Quantity, quality and reliability of the water supply.
(c) Water treatment and pumping equipment.
(d) Maintenance and operation of the water injection installations and possibly the monitoring of areal sweep using tracers.

27.1. Injection well completions

The completion of injection wells involves:

(a) The initial completion.
(b) The selective plugging of thief zones.

27.11. Initial completion

There are two possibilities: newly drilled injection wells or converted production wells.

A. Completion of new injection wells

The drilling and casing programme must provide for 7″ casing to the top of the reservoir.

The formation should be continuously cored then stimulated in open hole in order to maximise the injectivity.

The injection string is usually equipped with a safety joint.

B. Conversion of existing production wells

The production completion is first pulled out. If the casing is still in good condition and can support the expected pressures, injection can take place by way of the casing. If not, tubing must be run, with a packer just above the reservoir.

If the casing shoe is just above the reservoir the cement bond should be tested and if necessary remedial cementation carried out. If the well is cased over the entire reservoir the formation should be reperforated in order to ensure that the injection stands a good chance of succeeding.

To the same end, the above operations are followed by acid or solvent washing of the formation around the wellbore.

27.12. Detection and selective plugging of thief zones

It is often advisable to locate and then plug off the most permeable zones ("thief zones").

A. Detection of thief zones

Thief zones may be detected either by the use of a continuous flowmeter (see the "Logging course") or by adding a radioactive tracer to the water and recording a radioactivity profile.

The first method is applicable in cased hole and the second in open hole completions.

An example of the use of the continuous flowmeter in injection wells can be found in the Schlumberger "Production Logging" booklet (February 1963, example 6, page 32). It is recommended to make two runs at different injection rates and to compare the results.

There are two types of radioactive method:

(1) Water dosed with a tracer is injected via the tubing and clean water injected via the tubing/casing annulus (see Fig. 27.121). A gamma ray tool is lowered in the tubing, the cable passing through a stuffing box at the wellhead.

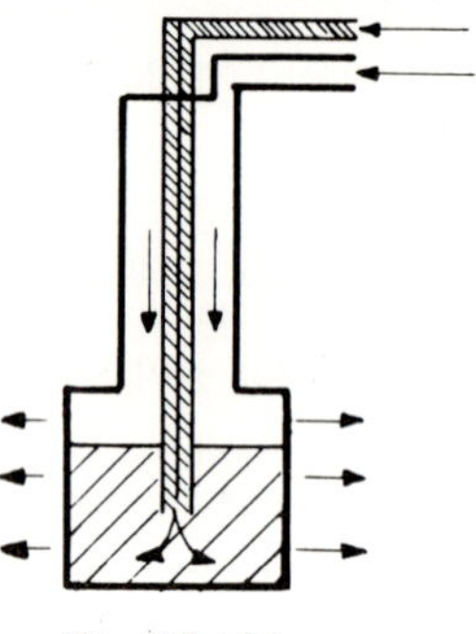

Fig. 27.121

The interface between the two waters is positioned at the top of the reservoir.

By simultaneous pumping the waters are injected into the formation. By slowly increasing the flow of clean water the interface is made to descend, its position being determined by the gamma ray tool. Knowing the injection rates and the position of the interface for each injection rate ratio, the injectivity profile of the formation can be calculated.

(2) The tracer is contained in plastic beads of which a certain quantity are mixed with the injected water. The more permeable the formation, the more beads become stuck on the sides of the hole at that level. A gamma ray tool is then run and the radioactivity profile thus obtained will be identical to the required injectivity profile.

B. Plugging agents

Amongst plugging agents in current use, many are dispersed solids or semi-solids in suspension (powders such as talc, zinc oxide, cement, colloids, resin emulsions soluble in caustic soda, etc.).

The technique used is a follows: the solids are injected in the form of particles whose size is such that they will only enter the pores of the most permeable parts of the formation.

Let us consider a solid particle which has entered a pore. It will travel into the formation until it reaches a point where it becomes trapped. The following particles will then back up behind it and the formation will become plugged at this point. The effect can be accentuated if a swelling material is used. The resulting plug may then be resistant to backflow.

In addition, the low permeability zones have a tendency to flow into the high permeability zones following a large reduction in injection pressure. This allows the low permeability zones to eliminate any plugging agent which may have been deposited there and helps to increase the plugging of the permeable zones.

The thief zones may also be plugged with cement or resin.

One service company is proposing selective plugging by the formation of precipitates in the rock, resulting from the in situ reaction between two fluids injected consecutively. The precipitates may be dissolved by hydrochloric acid in case unplugging is ever required.

27.2. Sources and treatment of injected water

The first investment required in a water injection project is for the drilling and completion of the injection wells. The second involves the range of installations required for the water supply, and for its treatment to enable it to pass through the reservoir without corroding the surface installations and well completions.

27.21. Sources of injected water

After injection has been in progress for a certain time, injected water appears at the production wells by way of the most permeable beds. The water is separated from the oil at surface and becomes available for re-injection into the reservoir. The volume of water available for recycling increases with time, since the oil production rates decrease as the water production rates increase.

It is of course always necessary to have an independent source of water, firstly to get injection underway and fill up the reservoir, and secondly to replace the volume of oil produced. Having two sources of injected water complicates its treatment since its composition changes with time. This requires a flexible treatment plant especially as mixing two different types of water at surface may result in chemical reactions whose undesirable effects must be eliminated before the water reaches the injection system.

1. Fresh water sources

(a) Surface waters held in rivers or lakes; they have the undesirable qualities of containing large amounts of oxygen, a high level of suspended solids (sand, animal or vegetable products and bacteria), and thus require a considerable amount of filtration equipment. In addition the composition of surface waters is variable.

(b) Alluvial waters tapped by shallow wells in the vicinity of surface water accumulations; they have the advantage of having been naturally filtered but are still subject to surface contamination and anaerobic bacteria may develop.

(c) Permeable shallow reservoirs are also very often fresh water bearing. However, their use is restricted to the provision of potable water in many areas.

2. Salt water sources

(a) In the vicinity of oil reservoirs there are often deep, salt water bearing formations. The water may be pumped to surface in specially drilled water supply wells, and in general requires the least extensive treatment. If the water contains no CO_2 or H_2S it should be protected against atmospheric contact and injected after filtration; the presence of CO_2 or H_2S requires special treatment. This type of water often contains anaerobic bacteria, particularly the sulphate reducing variety which must be positively eliminated.

(b) Finally, seawater may be used. It can be pumped from a submerged caisson. Seawater is usually corrosive and requires treatment to reduce its attack on metal equipment.

Since the mixing of different waters is unavoidable it is important that their compositions are well known. Special attention should be given to the presence of the following pairs of ions, since they may form precipitates:

(a) Barium and sulphate.
(b) Calcium and sulphate.
(c) Calcium and carbonate.
(d) Iron and sulphur.
(e) Iron and oxygen.

The precipitation of sulphates is the most damaging, since they are not soluble in present solvents or acids; the use of polyphosphates helps to avoid sulphate precipitation in the reservoir.

27.22. Water treatment

A. *Treatment objectives*

The objectives of water treatment are:

(a) To avoid plugging the reservoir.
(b) To avoid the corrosion of the injection system (surface and down-hole).
(c) To avoid the swelling of shales.

1. *Plugging*

This may be due to:

(a) Suspended solids.
(b) Corrosion.
(c) Bacteria (the most damaging being sulphate reducing bacteria).
(d) The incompatibility of the waters.

2. *Corrosion*

This must be avoided both for the protection of metallic equipment and in order to avoid the plugging referred to above. Corrosion is principally due to the presence of gas dissolved in the water: H_2S, CO_2 and oxygen, and to bacterial action.

3. *Swelling of shales*

The introduction of foreign water into an argillaceous reservoir may cause bentonitic shales to swell due to an exchange of ions between the water and the shale; this results in a reduction of the rock permeability. The swelling is pH sensitive, an acid water giving rise to a contraction of the shales, but such a water would be extremely corrosive.

The greatest risk of plugging is in the immediate vicinity of the wellbore. One method of treatment which has been successfully used is the injection of a large volume of hydrochloric acid into the formation, letting it remain there for several hours before the start of water injection. In the area in which it has been used, this treatment protects the shales from swelling when contacted by the injected water.

B. *Treatment methods*

Three methods of treatment are available: physical, chemical and biological.

Physical treatments include filtration, oil-water separation, settling and degassing, the last two being especially common in installations open to the atmosphere.

Chemical treatments include the addition of surface tension reducers and corrosion inhibitors.

Biological treatment involves the use of bactericides, bacteriostats and algicides.

The physical treatments will now be examined in more detail:

(a) Filtration. The basic types of filter are: standard sand packs (downward flow), special sand packs (upward flow), diatomaceous earth (the most efficient) and disposable cartridge filters.

(b) Oil-water separation. Depending on the percentage of water and the degree of separation required, dehydrators, free water separators or wash tanks may be used.

(c) Settling (normally in open systems) in tanks for the deposition of sediments.

(d) Elimination of dissolved gasses (O_2, H_2S and CO_2). There are two types of treatment:

- Stripping with natural or inert gas by countercurrent flow in a closed column through which the water passes slowly.
- Vacuum degassing.

27.23. Types of water injection systems

There are basically two types of water injection system: the closed type in which the water is protected against atmospheric contact and the open type involving treatment by aeration and sedimentation. The open type requires more attention than the closed type.

In a closed system there may be two independent circuits, a make-up water circuit and a recycled water circuit, and injectors may be designated for the injection of make-up or recycled water only.

27.3. Pumps

Two types of pumps may be used, reciprocating or centrifugal.

Reciprocating pumps have the dual advantage of high efficiency and flexibility. They are able to operate over a wide range of pressure and flowrate within their hydraulic power limit. Their principal disadvantage is their requirement for frequent maintenance, especially when corrosive or sand-bearing fluids are being pumped. They can be designed to operate at pressures above 200 bar or at rates of around 1 000 m^3/day.

Centrifugal pumps are less efficient that reciprocating pumps but are generally more robust and cost less to maintain.

They may be mounted horizontally or vertically and are generally used when rates of the order of several thousand m^3/day are required at pressures lower than 100 bar.

They are also used for pressures under 20 bar, at rates of some hundreds of m^3/day.

27.4. The Operation of a water injection system

The operation of a water injection system should not be left to individual initiative, it requires a comprehensive operating programme.

This programme should include:

(a) Water quality control (physical, chemical and bacteriological analyses).

(b) Corrosion control (using corrosion coupons, etc.).

(c) Regular inspection of surface equipment (for corrosion, leaks, operation of safety devices and automatic equipment).

(d) Regular inspection of the condition of injection and source wells.

The inspection of injection wells includes:

(a) Ensuring that there are no leaks in the completion. This is achieved by:

. Temperature surveys. Two temperature profiles must be recorded, one at static conditions and one during injection. The method is clearly explained in the Schlumberger paper: "Temperature logs in production and injection wells" by Loeb and Poupon (27th Meeting of the European Association of Exploration Geophysicists, Madrid, May 5-7, 1965).

. Flowmeter surveys.

. Radioactive tracer surveys.

A leak may occur in the injection tubing string, in the packer or in the casing. Well surveys and leak detection programmes are the responsibility of the petroleum engineering department.

(b) Regularly checking the annulus pressures, so as to avoid the casing being subjected to the injection pressure.

(c) Routinely recording the wellhead pressure and calculating the skin effect from pressure fall-off surveys in order to detect any increase in formation damage. In practice, if the wellhead pressure falls by rather more than the tubing friction losses during the first few seconds after injection has been stopped, an appreciable amount of damage can be assumed. This damage can be reduced by:

. Acidisation (*cf* Stimulation Course).

. Backflowing the well after stopping the injection.

. Corrosion and bacterial control.

(d) Checking the condition of the injection string using a tubing caliper log. If corrosion is found to be extensive the tubing should, if possible, be replaced before it bursts.

27.5. The Use of tracers to control sweep efficiency

In a limited area of a field, a small quantity of radioactive or chemical material is added to the injected water, a different substance being used for each injection well.

It is thus possible, when water breakthrough occurs at a production well, to determine from which injection well the water comes.

In large fields in which complete water tracing is required, the same tracer may be used for several sufficiently widely spaced injection wells.

Typical chemical tracers used are: formaldehyde, thiocyanates, nitrates, dichromates, iodides (Ref. 23), etc.

Tracers are not required to be continuously injected and, since the objective is to track the frontal advance, only the initial volume of injection water requires "tagging". Tracer injection need only be continued long enough to ensure that its concentration at the front will be adequate for identification at breakthrough.

In part of the Brea-Olinda field, in the United States (Ref. 23), water injection was followed by polymer injection. In both cases chemical tracers at a concentration of about 1 000 ppm were added to the injected water for 2 days. By comparing tracer arrival times the improvement in sweep efficiency due to the polymer injection was estimated.

APPENDIX 2.1

A study of the comparative merits of water and gas injection in the lower Gassi Touil reservoir (Ref. 4)

The lower Gassi Touil reservoir is a very unsymmetrical anticline with its axis running North-South, comprising three domes separated by two saddles, the largest saddle lying between the northern and central domes. The original reserves in place were:

Oil: 97.4×10^6 t Gas: 18.3×10^9 m^3 at standard conditions.

There are three gas-caps in the reservoir, the underlying oil being located on the eastern flank and completely filling the saddles.

At 1st January, 1969, the cumulative production was 12.8×10^6 t of stock tank oil, 13.2 % of the original reserves in place.

The predicted recovery factors for 1979, calculated by material balance, were: expansion drive: 28%, gas injection: 31.5%, water injection: 36%. However, these results were only regarded as approximate because of the rather simple method of calculation used. The study was repeated using the Franlab three-dimensional, three-phase "TRI-TRI" computer program and the following results obtained for the various cases studied:

	Pressure maintenance by gas injection	Water injection throughout, using Albian water	Water injection in the Saddles, using Albian water	Gas recycling from lst stage separator	Mixed injection
Additional Production (10^6 t)	2.8	3.2	3.0	2.6	3.4
Capital Investment (10^6 F)	64	53	45	48	58
Investment/ Production ratio (F/t)	22.8	16.6	15.0	18.4	17.1

APPENDIX 2.2

Practical interpretation of pressure fall-off curves

A. Introduction

Two cases must be considered, depending on whether injection was stopped before or after "fill-up". In practice it is often useful to obtain an estimate of injection well skin damage before fill-up is complete. If required, a well could then be stimulated to improve its injectivity without delay. When water injection is stopped one of the following may occur:

(a) The wellhead pressure falls slowly and the well remains full of water for a long time. This happens when the reservoir pressure is high. The small amount of after-injection which takes place is a result of the expansion of the fluid in the wellbore.

(b) The wellhead pressure drops rapidly to atmospheric soon after shut-in and the water level in the well falls. The after-injection corresponds to this fall in the water level.

Several authors (Refs. 17, 18) have proposed methods of interpreting pressure fall-off curves in injection wells, Kazemi et al. having defined a numerical model for use in the analysis of difficult cases.

For a liquid filled reservoir the classical methods of analysis can normally be used (plot of P_{ws} vs $\log (t - t_1)$ or P_{ws} vs $\log t/(t - t_1)$, t_1 being the time at which the well was shut in) but in some cases difficulty may arise, particularly where the total compressibility varies throughout the system. (For further details see Ref. 18).

For a partly filled reservoir, extrapolation of the plot of ΔP vs $\log [t/(t - t_1)]$ often leads to unrealistic (even negative) values of the static reservoir pressure. In these cases the analytical solutions obtained by Hazebroek *et al.* are extremely useful.

B. Analysis of pressure fall-off curves in reservoirs filled with fluids of equal mobility

Consider a reservoir in which water injection has taken place at a constant rate, Q, before shut-in. During the pressure fall-off we have:

$$P_{ws} = \overline{P} + \frac{\mu Q}{4\pi kh} \log [t/(t - t_1)] + \text{constant}$$

where $\overline{P}$ represents the average reservoir pressure.

The kh product and skin effect are calculated as for a pressure build-up, but the calculation for average pressure is somewhat different.

Consider a five spot pattern in a homogeneous reservoir of constant thickness.

The pressure along the 50% equipressure contour, line D in Fig. A.2.2.1, is equal to the average pressure of the system. We can thus approximate the behaviour of our system by representing the cross-hatched square in the figure by a circle of the same area, A:

$$\pi r_e^2 = A$$

Hazebroek, Rainbow and Matthews have shown that the average pressure $\overline{P}$ is related to the extrapolated pressure P^* on the plot of P_{ws} vs log $[t/(t - t_1)]$ for $\frac{t}{t - t_1} = 1$ by the equation:

$$\overline{P} - P^* = -\frac{\mu Q}{4\pi kh} E_i\left(-\frac{A}{4\pi k t_1}\right) \qquad \text{Eq. A.2.2.1.}$$

The value of the E_i function in the above equation may be obtained from the table given in Volume III of this production course (by P. Chaumet).

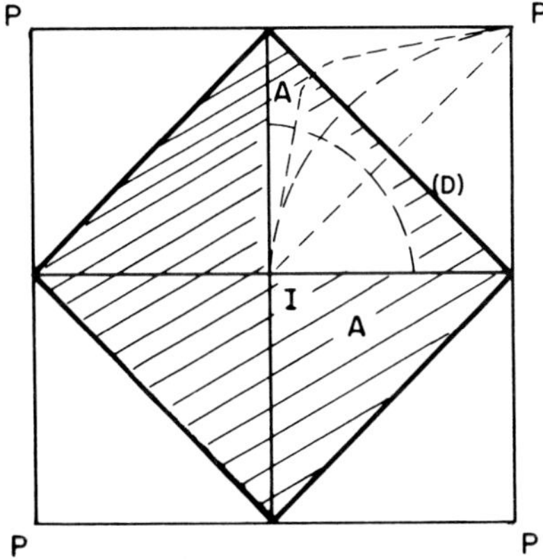

Fig. A.2.2.1.
(Ref. 17).

Example

The pressure recording shown opposite was obtained during a pressure fall-off survey in an injection well.

The well, reservoir and fluid characteristics are as follows:

$r_w = 10.5$ cm $\mu_w = 0.6$ cP $\phi = 16\%$

$W_i = 378{,}000\ m^3$ $B = 1.0\ m^3/m^3$ $c_o = c_w = 4.35 \times 10^{-5}\ bar^{-1}$

$Q = 226\ m^3/day$ $h = 15$ m $c_p = 5.8 \times 10^{-5}\ bar^{-1}$

The well is in a five spot pattern of area $1.61 \times 10^9\ cm^2$.

From Fig. A.2.2.2 the slope of the fall-off curve is $m = 8.95$ bar/cycle, from which a permeability of 21.8 mD and a skin effect of -3.78 can be calculated (the well had been fractured).

The diffusivity in the swept zone, assuming simultaneous flow of oil and water, is given by:

$$K = \frac{k}{\phi \mu c}$$

with

$$c = c_t = S_o c_o + S_w c_w + c_p$$

In this case $S_o = 0.2$, $S_w = 0.8$, and

$$K = \frac{21.8 \times 10^{-11}}{0.16 \times 0.6 \times 10^{-2} \times 1.05 \times 10^{-10}} = 2\,240 \text{ CGS}$$

The value of P^* obtained by extrapolating the fall-off curve, -22.2 bar, obviously has no physical significance. The average pressure $\overline{P}$ is calculated from equation Eq. A.2.2.1 with $A = 8.05 \times 10^8\ cm^2$ (area of the cross-hatched square in Fig. A.2.2.1, being half the pattern area) and

$$t_1 = \frac{378\,000}{226} = 1\,672.6 \text{ days} = 40{,}100 \text{ hours}$$

The argument of the E_i function is given by:

$$x = \frac{8.05 \times 10^8}{12.57 \times 2.24 \times 10^3 \times 4.01 \times 10^4 \times 3.6 \times 10^3} = 1.97 \times 10^4$$

and

$$-E_i(-x) \approx 7.91$$

Thus :

$$\overline{P} - P^* = 7.91 \frac{8.95}{2.302} = 30.7 \text{ bar}$$

$$\overline{P} = 30.7 - 22.2 = 8.5 \text{ bar}$$

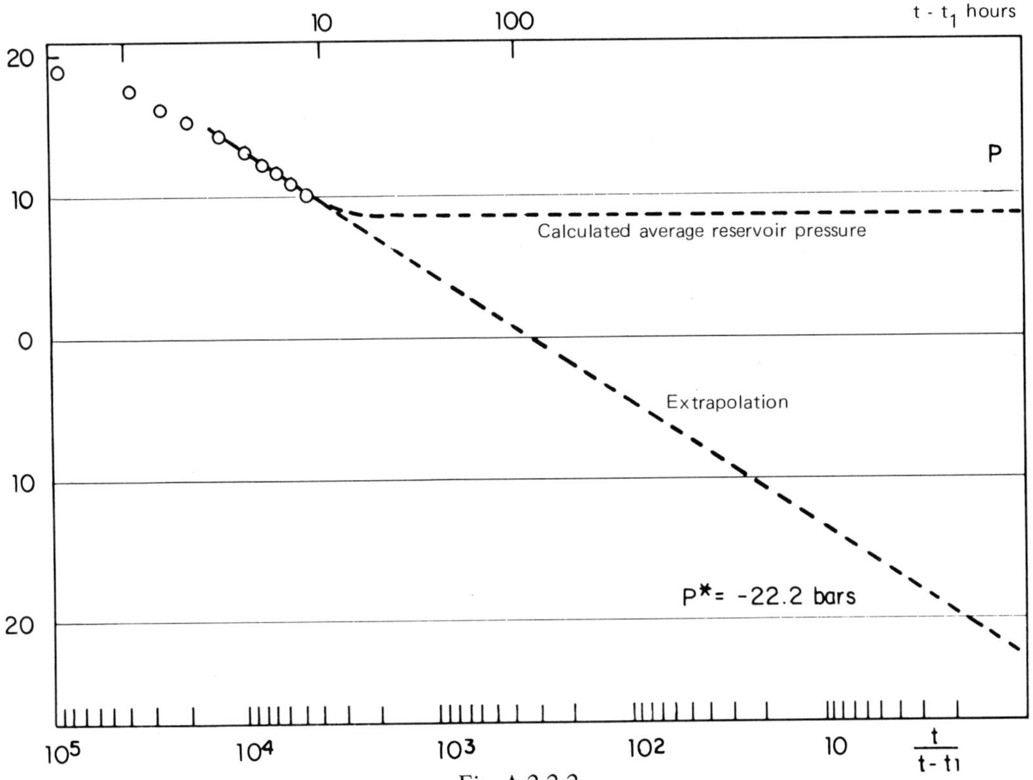

Fig. A.2.2.2.
(Ref. 17).

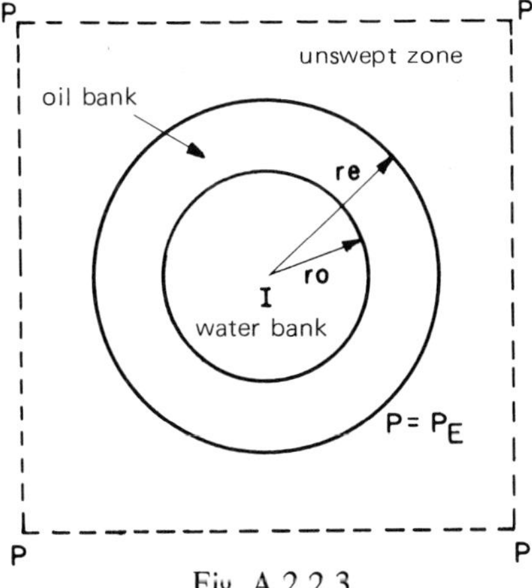

Fig. A.2.2.3.

C. Hazebroek's analytical solutions for a partly liquid filled reservoir

Hazebroek and others obtained analytical solutions for the pressure fall-off after stopping water injection in the following two cases :

Case No. 1. Water bank and oil bank of equal mobility with after-injection.
Case No. 2. Water bank and oil bank of different mobility with no after-injection.

Evidently these hypotheses do not cover every eventuality, and Hazebroek's solutions cannot be used in every case.

There will occasionally be the need for a numerical model of the type defined by Kazemi (Ref.18).

Figure A.2.2.3 shows the limits of the various zones around an injection well.

Let us assume that the water displaces the oil and gas until the gas saturation in both oil and water banks is the uniform residual value S_{gr} and the oil saturation in the water bank is the uniform residual value S_{or}. Only water flows in the water bank and only oil flows in the oil bank. This approximation is valid for oils of viscosity less than or equal to 50 cPo. As there are few injection projects involving oils of higher viscosity, the following analysis is of general application.

Let P_e be the pressure at the outer radius of the oilbank. Since the compressibility of the unswept zone is high (due to its high gas saturation) we can consider P_e to be the pseudo static pressure. It can then be shown that, whether or not there is after-injection, the bottom-hole pressure P_{ws} for large values of Δt can be written as:

$$P_{ws} = P_e + b \exp(-\beta \Delta t)$$

Thus there is a linear relationship between log $(P_{wf} - P_e)$ and Δt. P_e is determined by successive approximation.

In certain cases it is necessary to bracket the correct value of P_e (curve C) with values which are too high (curve A) and too low (curve B) (Fig. A.2.2.4).

The following units are used in the formulae developed for Cases 1 and 2 :

k mD
d_t tubing diameter cm
β (hours)$^{-1}$
Q m^3/hour
b bar
ρ g/cm^3
μ cP
h m

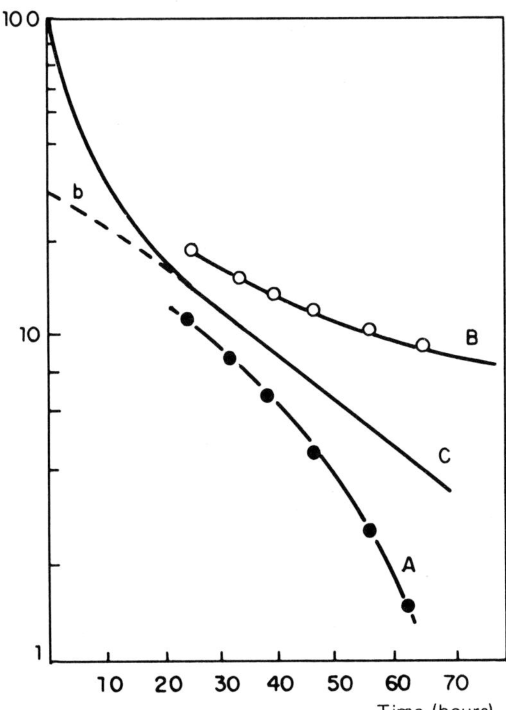

Fig. A.2.2.4.
(Ref. 17).

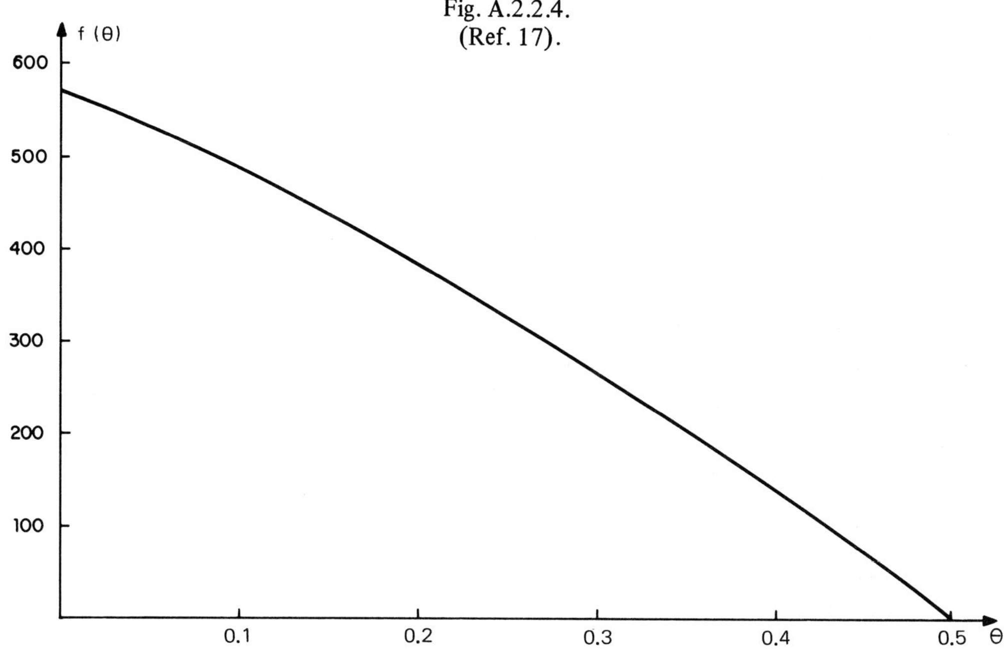

Fig. A.2.2.5.
(Ref. 17).

Case No. 1. The formation capacity is given by:

$$kh = \frac{Q\mu}{b} \frac{1 - C_1 - C_2}{(1 - C_3)} f(\theta)$$

where

$$\theta = \frac{C_1 (1 - C_3)}{2(1 - C_1 - C_2)}$$

and

$$C_3 = \frac{P_{ws} - P_e}{b} C_1$$

$f(\theta)$ is given in Fig. A.2.2.5.

The values of C_1 and C_2 depend on the behaviour of the wellhead pressure P_t (slow decline or sudden drop).

	Wellhead pressure fall-off after shut in	
	Slow decline	Sudden drop to atmospheric
C_1	$Ac_w (P_{ws} - P_t)$	A
C_2	0	$P_t C_1 / b$

where

$$A = 8 \times 10^{-4} \frac{d_t^2 \, b\beta}{\rho Q}$$

The skin factor is obtained in the usual way:

$$S + \mathrm{Log} \frac{r_e}{r_w} = \frac{2\pi kh}{Q\mu} (P_{ws} - P_e)$$

The external radius r_e being calculated from the cumulative volume of water injected, w_i:

$$r_e = \sqrt{\frac{w_i}{\pi \phi (S_g - S_{gr}) h}}$$

Case No. 2. In this case, with the same assumptions as made above, we have from Ref. 17:

$$kh = 6.34 \frac{Q\mu}{b} F$$

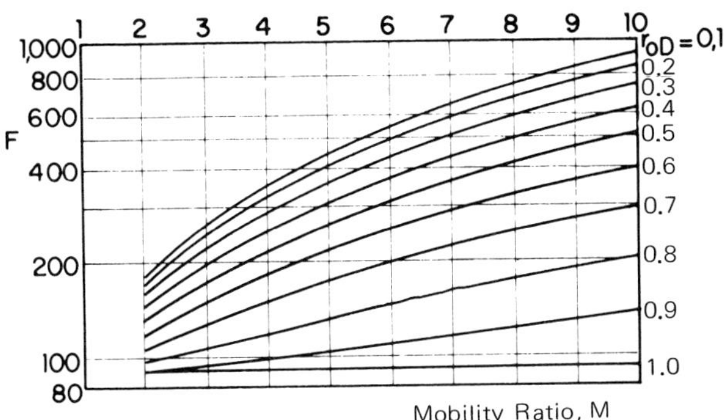

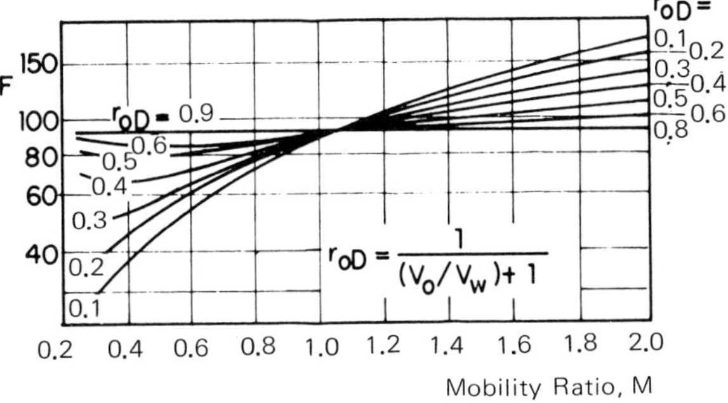

Parameter F for calculation of kh with $\alpha = 1$.

Fig. A.2.2.6.
(Ref. 17).

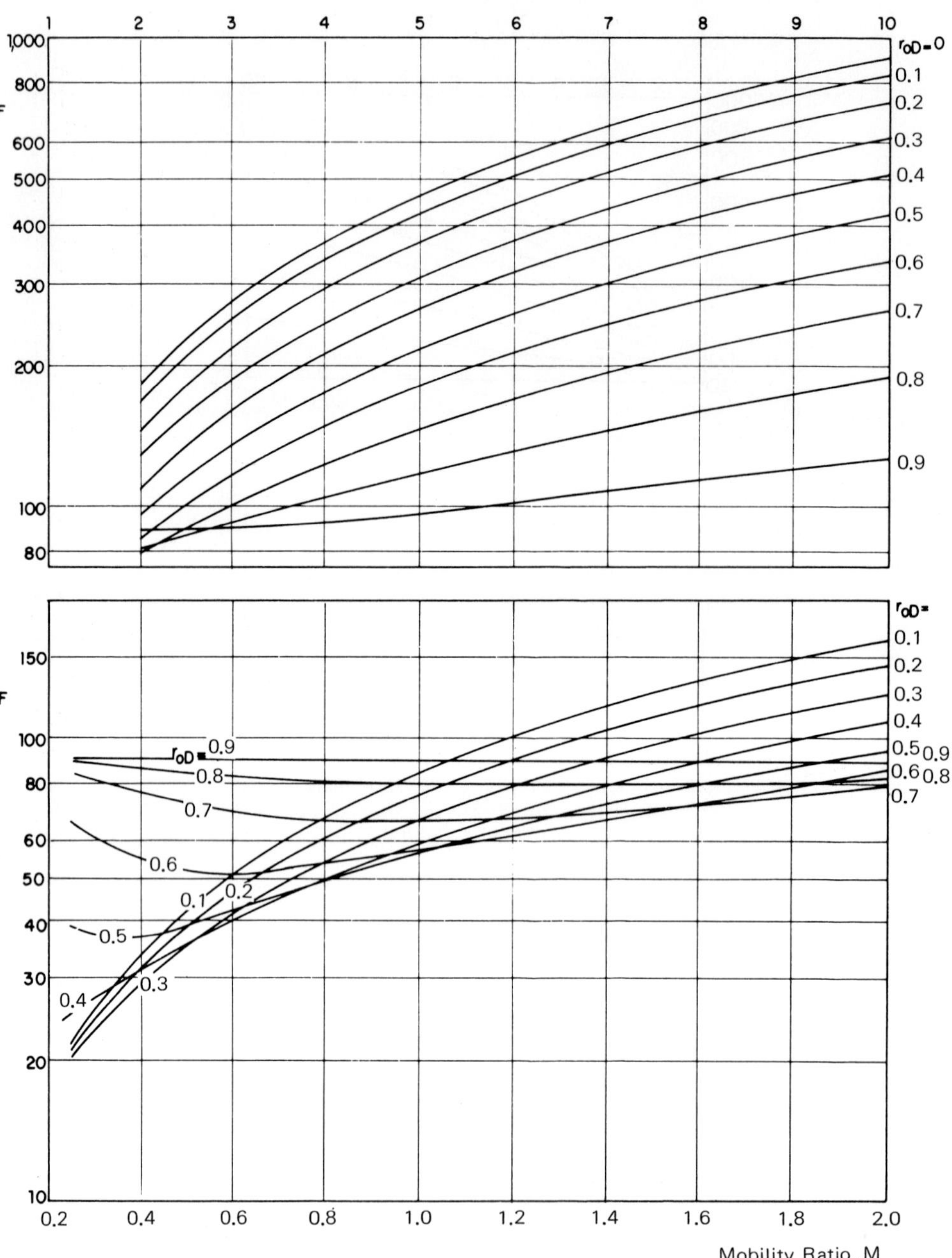

Parameter F for calculation of kh with $\alpha = 2$.

Fig. A.2.2.7.
(Ref. 17).

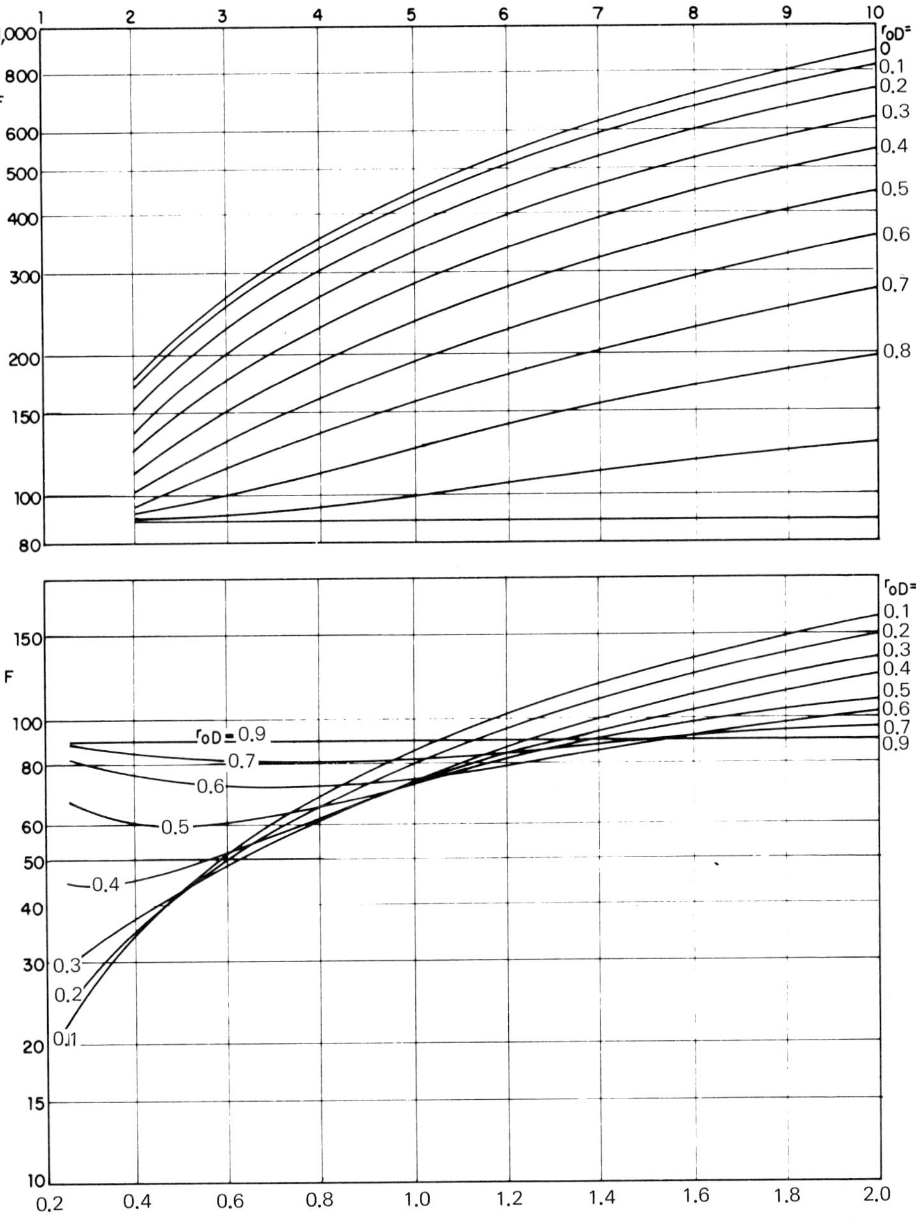

Parameter F for calculation of kh with $\alpha = 4$.

Fig. A.2.2.8.
(Ref. 17).

where F is an expression involving Bessel functions and is dependent on the following factors:

(a) Ratio of the total compressibility of the oil bank to that of the water bank, α.

(b) Mobility ratio, M.

(c) Reduced radius r_{0D}: ratio of the outer radius of the water bank to the outer radius of the oil bank:

$$r_{0D} = \frac{r_0}{r_e} = \sqrt{\frac{S_g - S_{gr}}{S_0 - S_{0r} + S_g - S_{gr}}}$$

A set of curves of F are given in Figs. A.2.2.6. to A.2.2.8.

The skin effect can be estimated from the following equation:

$$P_{ws} = \frac{Q\mu}{2\pi kh}\left[S + \text{Log}\frac{r_e}{r_w} - (M - 1)\ \text{Log}\ r_{0D}\right] + P_e$$

which in practical units becomes:

$$S + \text{Log}\frac{r_e}{r_w} = 2.26 \times 10^{-3}\ \frac{(P_{ws} - P_e)\,kh}{Q\mu} + (M - 1)\ \text{Log}\ r_{0D}$$

REFERENCES

1 KASTROP, J.E., "Oil recovery technology : key to future US supply". *Petr. Engineer.*, December 1972.

2 MARIOTINI, P. and de SAINT PALAIS J., "Le maintien de pression par injection d'eau sur le champ de Zarzaitine. Reservoir F4". *Revue de l'AFTP,* **190**, juillet-aout 1968.

3 BINEAU, M., "Récupération secondaire dans les zones peu perméables de Parentis". *Revue de l'AFTP,* **198**, novembre-décembre 1969, p. 83-89.

4 RIBUOT, M., "Etude du maintien de pression du gisement de Gassi-Touil inférieur". *Revue de l'AFTP*, **198**, novembre-décembre 1969, p. 90-108.

5 SMITH, CR., *Mechanics of secondary oil recovery.* Reinhold Publishing Corporation, New York, 1966.

6 CAUDLE, Ben H., "Fundamentals of reservoir engineering. Part II". Video tape course, SPE, 1968.

7 IFP Division Production, "Phénomènes physiques et chimiques intervenant dans l'exploitation des gisements d'hydrocarbures", Vol. I. Rapport IFP, réf. 18 734, novembre 1970.

8 CHAKIB KHELIL, "A correlation of optimum free gas saturation with rock and fluid properties". SPE Paper n° 1983.

9 HOMGREN, C.R. and MORSE, R.A., "Effect of free gas saturation on oil recovery by water flooding". *Trans. AIME,* 1951, **192**, p. 135.

10 DYES, A.B., "Production of water drive reservoirs below their bubble point". *JPT*, October 1954, 6, p. 31.

11 KENNEDY, H.T. and GUERRERO, E.Y., "The effect of surface and interfacial tensions on the recovery of oil by water flooding". *JPT* May 1954, **201**, p. 124.

12 BASS Jr., D.M., "The effect on oil recovery of waterflooding at pressures above and below the bubble point". M S thesis, Texas A and MU, College station, January 1955.

13 WALTON, D.L., "A comparison of the performance of waterfloods using similar refined and crude oils". M S thesis, Texas A and MU College station, August 1959.

14 KYTE, J.R., STANCLIFT Jr., R.J., STEPHAN Jr., S.C. and RAPOPORT, L.A., "Mechanism of waterflooding in the presence of free gas". *JPT* September 1956, 8, p. 215.

15 JACKSON, M.L., "Oil recovery by combination of water and miscible fluid injection in radial systems". M S thesis, U. of Oklahoma, Norman, 1957.

16 COLE F.C., *Reservoir Engineering manual.* Gulf Publishing Co., Houston, 1961, 104, 6.

17 HAZEBROEK, P. et coll., "Pressure fall-off in water injection wells". *Trans. AIME,* 1958, 213, 250-260.

18 KAZEMI et coll., "Problems in interpretation of pressure fall-off tests in reservoirs with and without fluid banks". *JPT,* sept. 1972.

19 MUSKAT, M., *Physical principles of oil production.* Mc Graw-Hill Book Co., Inc., New-York, 1949.

20 BURWELL, E.L., "Multiple tracers establish waterflood flow behavior". *Oil and Gas Journal,* 28 nov. 1965 ; *Oil and Gas Journal,* vol. 64, n° 48, p. 76-79, 28 nov. 1966.

21 KELLDORF, W.F., "Radioactive tracer surveying, a comprehensive report". *JPT,* June 1970.

22 CALHOUN, T.G. and HURFORD, GT., "Case history of radioactive tracers and techniques in Fairway Field". *JPT,* October 1970.

23 SHERBORNE, J.E., SAREM, A.M. and SANDIFORD, B.B., "Flooding oil-containing formations with solutions of polymer in water". *7th World Petroleum Congress,* Proceedings, Elsevier Ping CY, p. 509.

3

gas injection in an oil reservoir (immiscible displacement)

31. INTRODUCTION

Gas injection in an oil reservoir takes place either into a gas-cap if one exists, or directly into the oil zone.

The injected gas is practically always of a hydrocarbon base. Air injection has been attempted, but has many disadvantages (well corrosion, oil oxidisation, risk of explosion, etc.).

A. Gas injection into a gas-cap

When a gas-cap originally exists in a reservoir, or when one has been formed by segregation during primary production, gas injection helps to maintain the reservoir pressure while forcing gas into the oil zone and driving the oil towards the production wells. This process is analogous to the rise of the oil water contact when water is injected into an underlying aquifer.

B. Gas injection into an oil zone

When gas injection takes place in a reservoir without a gas-cap the injected gas flows radially from the injection wells, driving the oil towards the production wells.

The principal factor involved in the decision to commence gas injection is the availability of a nearby source of cheap gas in sufficient quantities. The recycling of produced gas is a major source, but can only slow down the reservoir pressure decline, not halt it. Secondary gas must be obtained either from an adjacent gas reservoir or from a nearby gas pipeline.

It is quite simple to calculate the quantity of gas required for complete pressure maintenance.

Consider an oil reservoir producing at a gas/oil ratio (GOR) of R, such that: $R = R_s + R_c$ (the producing GOR is equal to the sum of the solution GOR and the excess GOR due to the circulation of free gas.

The production of a unit volume of stock tank oil corresponds to a withdrawal of a reservoir fluid volume of: $B_o + R_c B_g$, while the surface gas production is: $R_s + R_c$.

If I percent of the produced gas is re-injected, its volume at reservoir conditions will be: $I(Rs + Rc)Bg$.

Complete pressure maintenance is assured when the reservoir volumes produced and injected are equal, that is when:

$$I = \frac{B_o + R_c B_g}{(R_s + R_c) B_g} \qquad \text{(Eq.31.1)}$$

In the special case in which no free gas is produced we have:

$$I = \frac{B_o}{R_s B_g} \qquad \text{(Eq.31.2)}$$

For normal oils this value of I is greater than 1. Thus the calculation indicates the need for a source of make-up gas.

In Chapter 2 the relative advantages of gas and water injection were examined and the following conclusions drawn:

(a) The capital investment required for gas injection is usually higher than that required for water injection.

(b) The microscopic displacement efficiency of gas is much less than that of water.

It should be noted that if dry gas is injected into an oil reservoir, the produced oil is made up of both oil displaced from the porous medium and oil fractions vapourised by the injected gas. If the oil is very light, the mass of oil vapourised may be extremely high and a high oil recovery will result. This is the case for two reservoirs in the USA: Pickton field in East Texas and Raleigh field in Mississippi. Various methods for the calculation of oil vapourisation have been published in the literature (Refs. 6, 7, 8). In what follows, however, we shall assume that oil vapourisation is minimal in comparison with oil displacement.

There are certain conditions under which water injection should not be considered:

(a) A very extensive gas-cap may form a preferential path for injected water which will thus by-pass the oil zone. By comparison, gas injection into the gas cap may result in additional oil recovery for the price of a few gas injection wells.

(b) A reservoir with a high initial water saturation may not be suitable for water injection. There is a risk that no front will be formed and that oil and water will flow in parallel, giving a low recovery efficiency.

Under these conditions, oil recovery by gas recycling may be possible, as long as the free gas saturation in the reservoir is not too high.

In certain Pennsylvanian fields with low gas saturations, satisfactory recovery factors (economically speaking) have been obtained by gas injection in spite of initial water saturations of 40-60% and oil saturations of 25-35%.

If the reservoir has sufficiently high vertical permeability, gas-cap injection will result in higher recovery than injection into the oil zone. The area at the gas-oil contact is large whereas in radial displacement the area of contact between gas and oil is initially small. A large contact area reduces the risk of preferential channeling.

It is advantageous to start gas injection into the gas-cap as soon as possible. In this way the formation of a high free gas saturation in the oil zone and an increase in oil viscosity are avoided, thus allowing a high oil relative permeability and productivity to be maintained.

There are two other comments to be made:

(a) Every gas injection project should be preceded by laboratory experiments (displacement studies on cores, physical models, etc.).

(b) Gas injection for enhanced recovery may also be regarded as underground gas storage and this may render a project economically attractive in certain circumstances.

32. INJECTION WELL LOCATION

In general, the increase in recovery which gas injection can provide does not warrant the drilling of a large number of new injection wells, thus most of the injection wells are obtained by the conversion of existing producers.

A. Injection into the gas-cap

All the injection wells (converted producers or new wells) are concentrated around the top of the structure.

The converted production wells may need to have their existing perforations cemented off and new perforations made in the gas-cap.

The adequate number of injection wells depends on the total injection rate required. The capacity of each injection well can be estimated using an equation of the well-known form:

$$Q = C(P_{iw}^2 - P_e^2)^n$$

where

P_{iw} is the bottom-hole injection pressure, and

P_e is the pressure at the external limit of the gas zone.

B. Injection into the oil zone

In this case the well density varies widely, but there is always less than one injection well per production well, whereas this ratio is most common in water injection projects.

In the first gas injection projects there was no attempt to keep to a regular pattern, but nowadays an inverted seven-spot pattern is often selected (one injector + 6 producers, see Chapter 1). However, such a pattern is often difficult to achieve, due to the location of the wells drilled during the primary production phase.

The sweep efficiency may be obtained from the curves given in Chapter 1, but since the mobility ratio M is much greater than unity the estimates obtained are often suspect. It is safer either to use a mathematical or physical model, to base the calculations on the results of neighbouring fields or to install a pilot injection scheme and extrapolate from its results.

33. SWEEP EFFICIENCY

Both gas and water injection result in the formation of a displacement front, but for gas injection the front is less distinct (less variation in oil saturation). The injected gas does not wet the rock surfaces but sweeps through the oil and tends to form a continuous gas phase throughout the reservoir.

This happens very rapidly, because the critical gas saturation is low.

During injection into the gas-cap there is more chance that a distinct front will be maintained, since gravity will assist in the segregation of the gas and liquid phases.

Yuster and Day performed some gas displacement experiments on core samples at pressures close to atmospheric. They obtained the following results:

(a) The cumulative oil production is directly proportional to the log of cumulative gas injected.

(b) For a given volume of gas injected, the greater the applied pressure gradient (i.e. the greater the gas velocity) the greater the volume of oil produced.

(c) For a given volume of gas injected, the greater the reservoir oil viscosity the lower the volume of oil produced.

For a given pressure gradient, the rate of oil drainage is slightly lower when there is an initial water saturation than when the rock is completely oil saturated (see Fig. 33.1).

On the other hand, for a given injected gas volume the recovery factor is greater when there is an initial water saturation.

For example, let us calculate the recovery factor r when 1 000 pore volumes of gas have been injected.

For the core initially oil-saturated: $S_{oi} = 1.00$, $(S_o)_{1\,000} = 0.53$.

$$r = 1 - \frac{0.53}{1} = 0.47$$

For the core with an initial water saturation: $S_{oi} = 0.5$, $S_{wi} = 0.4$, $S_{gi} = 0.1$, $(S_o)_{1\,000} = 0.17$.

$$r = \frac{0.5 - 0.17}{0.5} = 0.66$$

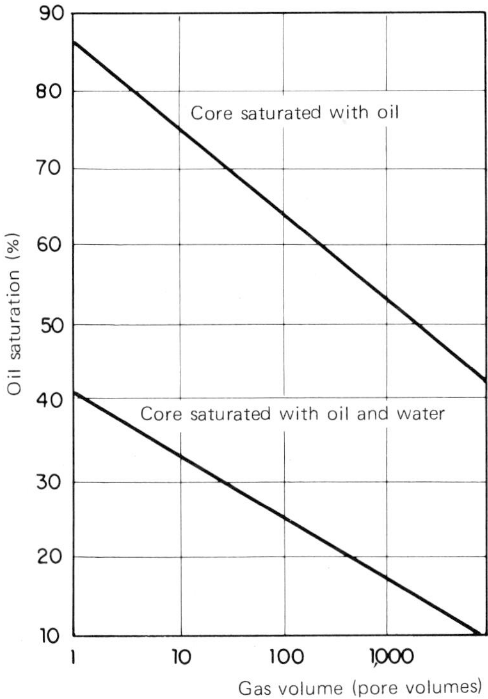

Fig. 33.1. The effect of water saturation on the drainage behaviour of an 800 md core plug.

From their experimental work, Yuster and Day suggested the following equation for oil saturation during gas injection:

$$S_o = S_{o1} - \alpha \operatorname{Log} V$$

where

S_o is the oil saturation when a gas volume V has been injected,
S_{o1} is the oil saturation when one pore volume of gas has been injected,
V is the volume of gas injected, expressed in reservoir pore volumes at the average reservoir pressure.

Using some rather less rigorous reasoning, the same authors arrive at the following equation:

$$\frac{1}{Q_o} = A + Bt$$

i.e. the inverse of the oil rate is directly proportional to time. Some field results have appeared to fit this type of equation, and in these cases the extrapolation of $\frac{1}{Q_o} = f(t)$ can be used to estimate the duration of economic production rates.

34. PRELIMINARY STUDIES AND FIELD EVALUATION OF INJECTION EFFICIENCY

In order to evaluate the efficiency of gas injection a radioactive tracer can be used to physically monitor the progress of the injected gas.

At the same time, calculations based on frontal displacement theory and material balance can be performed, which will provide a first estimate of the recovery efficiency.

34.1. The monitoring of sweep efficiency using radioactive tracers

Using different radioactive gases it is possible to "tag" the gas injected into each well in a limited area of a field. At the production wells (Fig. 34.11) the produced gas can be analysed and the tracers identified, thus determining at which well or wells the gas was injected([1]).

([1]) One difficulty is the detection and analysis of the tracers, because they appear at the production wells in extremely low concentrations.

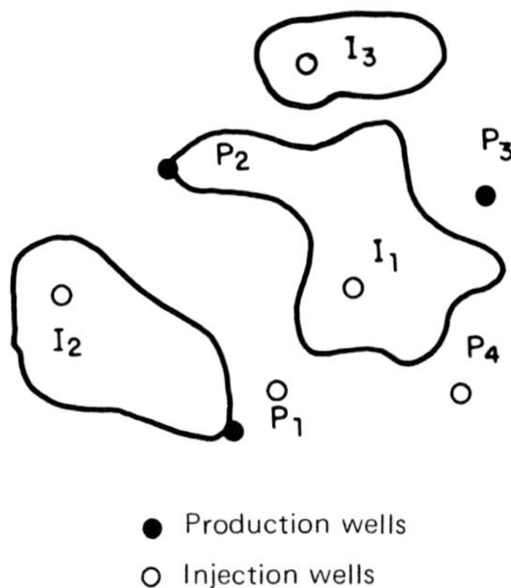

Fig. 34.11.

This enables a picture of the heterogeneity and anisotropy of the reservoir to be built up, and for the shape of the gas fronts to be estimated at given times.

As in the case of the tagging of injected water (see Section 27.5), continuous tracer injection is not required; only an initial volume of gas need be tagged. The tracer injection should only be continued long enough that it does not suffer from over-dilution during its displacement.

One of the fields in which tracers have been injected is the Hassi-Messaoud field in Algeria, where hydrocarbons irradiated to Tritium have been used with success. The first four alcanes were used (methane, ethane, propane and butane) and tracers were injected for about 80 days at each well. The tracers were contained in glass ampoules which were broken into a positive displacement pump mounted on the gas injection line at the well head. Samples, taken under pressure at the production wellheads, were then analysed. The methane, ethane, propane and butane were separated and passed through a detector able to measure the very weak β radiation from the irradiated hydrocarbon.

In other fields radioactive ethyl iodide has been used.

34.2. Calculations based on frontal displacement theory and material balance

The Buckley-Leverett theory is equally applicable to both gas and water injection. However, for the vertical flow of gas and oil it is not possible to neglect the effects of gravity.

Thus different equations must be used for the fractional flow of gas f_g depending on whether the injection takes place in the oil zone (horizontal flow assumed) or in the gas-cap (vertical flow assumed).

• Injection in the gas cap:

$$f_g = \frac{1 - \dfrac{k k_{ro}}{u \mu_o} (\rho_o - \rho_g) g}{1 + \dfrac{k_{ro}}{k_{rg}} \dfrac{\mu_g}{\mu_o}} \qquad \text{(Eq. 34.21)}$$

using the usual notation. Note that u is the volumetric gas injection rate per unit area, where the area is the total gas-oil contact area (pores plus rock).

• Injection in the oil zone:

$$f_g = \frac{1}{1 + \dfrac{k_{ro}}{k_{rg}} \dfrac{\mu_g}{\mu_o}} \qquad \text{(Eq. 34.22)}$$

The oil recovery Δr between time j (reservoir pressure $= P_j$) and $j + 1$ (reservoir pressure $= P_{j+1}$), assuming that there is no water encroachment, may be expressed as:

$$\Delta r = \frac{(1 - r_j)\,\Delta\left(\dfrac{B_o}{B_g} - R_s\right) - (1 + m)\, B_{oB}\, \Delta(1/B_g)}{\dfrac{B_o}{B_g} - R_s + \bar{R}(1 - I)} \qquad \text{(Eq. 34.23)}$$

where

m is the ratio of the initial gas cap volume to the initial oil zone volume (for injection into an oil zone with no gas-cap, $m = 0$),

I is the fraction of produced gas re-injected,

$\bar{R}$ is the average GOR between time j and $j + 1$,

the subscript B represents initial conditions (the liquid phase is assumed to be initially saturated).

From Eq. 34.23 we can derive the rate of re-injection required for complete pressure maintenance. When the pressure remains constant, the numerator of Eq. 34.23 is zero and, since the recovery is not zero, the denominator of the equation must also be zero and the equation is indeterminate. We thus have:

$$I = \frac{B_o - R_s B_g + R B_g}{R B_g} = \frac{B_o + R_c B_g}{(R_s + R_c)\, B_g}$$

For both gas-cap injection and oil zone injection the first step is to construct the curve $f_g\,(S_g)$, taking the viscosities at the saturation pressure.

For the case of complete pressure maintenance, the producing GOR and recovery can be calculated from this curve alone: the equation for Δr (Eq. 34.23) being of the indeterminate form 0/0.

In other cases the method of calculation depends on the type of injection.

A. Gas-cap injection

Let us examine the oil zone, which is invaded by the gas-cap due to the production of oil. We shall assume that any gas which comes out of solution does not join the gas-cap (no gas-oil segregation) and that the production wells are perforated such that they do not produce gas-cap gas.

Let us assume, as is often the case, that $S_{gi} = 0$ (there is no initial free gas in the oil zone). The tangent to the curve $f_g(S_g)$ is taken from the origin, and the abscissa of the point of intersection of the tangent with the line $f_g = 1$ gives us an estimate of the gas saturation in the part of the oil zone swept by the gas-cap; let this saturation be S_g.

For any fraction of gas re-injected I, we can see that the producing gas-oil ratio R at each pressure P is exactly that which would be obtained if there was no gas-cap and the oil was produced by solution gas drive. This is a consequence of the assumption that no gas segregation takes place.

We therefore:

(a) Calculate the oil production by solution gas drive and obtain R.
(b) Calculate Δr using Eq. 34.23.
(c) Calculate the new position of the gas oil contact.

For the last calculation a curve of reservoir volume as a function of depth is required, so that the increments of gas-cap volume can be converted into depths of the gas-oil contact. Note that, in the part of the oil zone invaded by the expanding gas-cap, the gas saturation only changes from S_{gi} to S_g. Thus when the gas-cap volume increases by ΔV at reservoir conditions, it invades a volume $\dfrac{\Delta V}{S_g - S_{gi}}$ of the oil zone.

The increase in gas-cap volume is the sum of the following two terms:

(a) The expansion of the initial gas-cap gas in place:

$$mN \frac{B_{oB}}{B_{gB}} (B_g - B_{gB})$$

(b) The cumulative gas injected: $(\Sigma\, \bar{R}NI\, \Delta r)\, B_g$.

B. Oil zone injection; reservoir with no gas cap: $S_{gi} = 0$

The curve $f_g(S_g)$ is drawn using equation Eq. 34.22. The tangent from the origin is constructed and other tangents drawn including that which has as its point of contact the value of f_g corresponding to the economic limit (e.g. $f_g = 0.9$). For each tangent the intersection S_g with the line $f_g = 1$ is read off and for each value the following calculations are performed:

Recovery:

$$r = 1 - \frac{B_{oB}}{B_o}\left(1 - \frac{S_g}{1 - S_w}\right)$$

where

S_w is the connate water saturation;

Producing GOR:

$$R = R_s + \frac{B_o}{B_g}\frac{f_g}{1 - f_g}$$

Thus we now have estimates of the final recovery and of the evolution of the GOR with cumulative production. It remains to relate the recovery to the reservoir pressure.

This can be done using the material balance equation:

$$\Delta r = \frac{(1 - r_j)\,\Delta\left(\frac{B_o}{B_g} - R_s\right) - B_{oB}\,\Delta\left(\frac{1}{B_g}\right)}{\left(\frac{B_o}{B_g} - R_s\right) + \bar{R}(1 - I)}$$

with

$$1 - S_g = S_\ell = S_w + (1 - S_w)\frac{B_o}{B_{oB}}(1 - r)$$

and

$$R = R_s + \frac{B_o}{B_g}\frac{k_g}{k_o}\frac{\mu_o}{\mu_g} = R_s + R_c$$

Note

When the injected gas does not come into contact with all the oil in the reservoir, the previous equations should be modified. A conformance factor e is introduced to represent the fraction of the reservoir contacted by the injected gas:

$$\Delta r = (1 - e)\,\Delta r_d + e\Delta r_e$$

$$\Delta r = \frac{(1 - r_j)\,\Delta\left(\frac{B_o}{B_g} - R_s\right) - B_{oB}\Delta\left(\frac{1}{B_g}\right)}{\left(\frac{B_o}{B_g} - R_s\right) + \bar{R}_e(1 - I)}$$

With

$$R_e = R_s + \frac{B_o}{B_g}\frac{\mu_o}{\mu_g}\left(\frac{k_g}{k_o}\right)_e$$

and

$$S_\ell = S_w + S_o = S_w + (1 - S_w)(1 - r_e)\frac{B_o}{B_{oB}}$$

where

Δr_e is the fraction of the original oil in place produced from the section of the reservoir contacted by injected gas as the reservoir pressure delines from P_j to P_{j+1},

Δr_d is the fraction of the original oil in place produced from the section of the reservoir not contacted by the injected gas,

$\left(\frac{k_g}{k_o}\right)_e$ represents the simultaneous flow properties of the oil and gas in the parts of the reservoir contacted by the injected gas.

All these calculations can be effectively performed by a computer. The program can be written in such a way as to also enable solution gas drive performance to be calculated ($e = 1, I = 0$).

Indeed, it is the comparison between recovery by gas injection and by solution gas drive which determines the attractiveness of such an enhanced recovery project.

Using a suitable computer program various runs can be made. Reservoir performance can be calculated for different injection start-up times, fractions reinjected, etc., and the results can be analysed for the optimum solution.

Obviously, with a conformance factor of less than unity the benefits of gas injection will be reduced. However, the project may still be justified if there is no market for the gas or if the field is to be used for underground gas storage.

35. INJECTION WELL COMPLETIONS

In the case of the conversion of existing production wells, the old completion must first be pulled out and the well cleaned at the reservoir interval by scraping or chemical washing. The condition of the casing must be checked and any leaks repaired.

The type of injection (gas-cap or oil zone) has a direct influence on the completion interval. In the first case, the oil producing interval is plugged off and the casing perforated at gas-cap level. In the second case it is not normally necessary to change the completion interval, unless certain zones are required to be eliminated because their high permeability to gas would cause "short circuiting". If the well is cased, cement squeezes or packers can be used to limit the injection zone. If not, it may be necessary to cement a liner in place which can then be perforated where required.

Injection may take place by way of the casing, or through tubing anchored with a packer just above the reservoir. It should be noted that wellhead pressures are likely to be much higher during gas injection than they were during oil production, and the wellheads may need to be changed to a higher series. For example, at Hassi-Messaoud the oil production wellheads are series 5 000 psi and the gas injection wellheads are series 10,000 psi.

When injection into several separate reservoirs in the same well is required, multiple completions may be used.

It is advisable not to select for conversion a production well which produced at a high water cut (a sign of the proximity of the aquifer, into which it is completely useless to inject gas).

36. PRODUCTION WELL COMPLETIONS

For horizontal gas-oil displacements, the original completion used during primary production is generally retained. During the course of the project however, it may become necessary to shut off zones through which gas is channeling preferentially (zones with a high GOR).

When injecting into the gas-cap, the upper limit of the production interval will need to be reduced as the gas-oil contact descends. This favours cased hole completions over open hole, since it is relatively simple to run packers to exclude gas production.

Excess gas production may be due to the phenomenon of coning. In this case the solution is to reduce the production rate of the well.

37. SURFACE INSTALLATIONS. COMPRESSION AND TREATMENT

Injection gas sources (separator gas, gas from nearby fields, inert or flue gases) may contain the following impurities:

(a) Hydrogen sulphide.
(b) Carbon dioxide.
(c) Oxygen.
(d) Water vapour.

H_2S, CO_2 and O_2 may cause corrosion of the surface pipework and downhole equipment, especially in the presence of traces of water. In addition to the effect on the installations, precipitates are formed which may damage (plug off) the formation.

The presence of water vapour, under certain conditions of temperature and pressure, may cause the formation of hydrates which can block the pipework.

37.1. Treatment methods

A. Desulphurisation

The gas is passed through absorption columns, where it comes into intimate contact with a chemical solution which can later be regenerated.

Reagents used:

(a) Sodium carbonate solution (regeneration by air current).
(b) Sodium phenolate (regeneration by heating).
(c) Amines (regeneration by heating).

B. Dehydration

Various dessicants are used, both solids (silica gel, activated aluminium, calcium sulphate, anhydrite, fluorite, etc.) and liquids (glycols).

There is practically no economic method for the removal of oxygen from gas. All possible precautions should be taken to avoid its accidental introduction into the surface pipework.

C. Filtration

Injection gas must be free from solid or liquid particles. Scrubbers and filters are thus installed in the system so as to remove all particles larger than a few microns. This requirement for a very high level of filtration is imposed by the porous medium, whose pores have an average diameter measured in microns (see Volume I). Filters are often situated both at the treatment centre and at each injection well.

37.2. Compressor installations

In most cases the available gas has to be compressed in order to be injected into the formation, and a central compressor station will be installed.

For a gas injection project the compressor power required can be calculated from the forecasts of gas pressure and flowrate.

From the start of gas injection to its abandonment, a progressively higher injection pressure will be required (the distance covered by the gas in the porous medium continually increases thus the pressure losses increase). The gas injection rate must increase after breakthrough as the producing GOR increases.

The installed compressor power for gas injection varies widely, from hundreds of kilowatts to tens of thousands of kilowatts. At Hassi-Messaoud the installed compressor power is one of the highest in the world (Hassi-Messaoud northern injection station: 15,500 kW, southern injection station: 20,600 kW).

Even in cases where one compressor could meet the power requirements, it is preferable to install several smaller compressors in order to provide flexibility and back-up in the case of mechanical breakdown. Most frequently, reciprocating compressors driven by gas engines are used.

Gas compression is neither isothermal nor perfectly adiabatic. Thus the gas law is written in the form: PV^n = constant, where n lies between 1 and $\gamma \left(\gamma = \frac{C_p}{C_v}\right)$. A value of 1.17 is often used for n.

The stage compression ratio cannot be greater than 5. Several stages of compression in series are thus often required to bring the source gas pressure up to the injection pressure.

The gas engines which drive the compressors may consume between 0.27 and 0.46 m^3/kWh according to American data.

During compression the gas temperature increases and cooling is subsequently required. This may be carried out:

(a) After each stage of compression.
(b) After the last stage only.
(c) If the final stage compression ratio is low, after each stage but the last.

Air coolers (e.g. the "aéroréfrigérants" at Hassi-Messaoud) or water coolers may be used. The surface installations are thus made up of:

(a) The treatment columns, if impurities or water vapour are to be removed from the gas. They are installed in pairs, each one being alternatively in service and under regeneration.
(b) The compressors.
(c) The coolers.

The gas is then distributed to the injection wells *via* a pipeline network fitted with the necessary pressure gauges, flowmeters and valves.

38. SPECIAL APPLICATIONS OF GAS INJECTION

38.1. The formation of a secondary gas-cap

In regions of complex geology, "attic" oil may become trapped at the top of an inclined formation, beyond the reach of the existing wells. One way of recovering this oil is to inject gas into the lower part of the reservoir. If the permeability and dip of the reservoir are great enough the gas will migrate to the top of the structure and form a secondary gas-cap. The attic oil is thus displaced down-dip and may be recovered by the same wells used to inject the gas.

38.2. Combined gas and water injection

The alternate or simultaneous injection of gas and water has been tested in the field (by Continental Oil in the USA, Sonatrach in Algeria).

The theory behind the process is as follows: it is thought that by successively injecting slugs of water and gas, a homogeneous mixture will be formed within the pores; due to relative permeability effects this mixture will behave as a fluid of low mobility; thus the mobility ratio of the system gas + water/oil will be reduced and the displacement efficiency improved.

Alternate injection is preferred to simultaneous injection for the following reasons:

(a) Higher injectivity,

(b) Cheaper and simpler surface equipement.

(c) Better vertical distribution of the two fluids throughout the thickness of the formation.

In a reservoir rock with intergranular porosity, the injected gas produces a trapped gas saturation at the centre of the pores which leads to a reduction in the residual oil saturation. This technique should be called "displacement by water with gas injection" since its most effective application is probably when water is the predominant displacing fluid.

It should be noted that as early as 1966, the *Société Nationale Repal* was conducting tests of alternate water and gas injection in the Hassi-Messaoud field. These tests have not yet proved very conclusive. A large decrease in permeability to gas was observed as soon as the first slug of water had been injected.

The advantages of this method of injection have to be evaluated in each particular case; it is not a method of universal application, and despite the increased recovery which may be predicted, the problems involved in putting it into practice should not be under-estimated.

38.3. Foam injection

Immiscible gas displacement presents a certain number of disadvantages, and the efficiency factors E_s ,E_d and E_i are low. Foam injection is a derivative of gas injection, and has been proposed because of its higher efficiency.

We shall study this process in detail in Chapter 7.

REFERENCES

1 POTTIER, J., DELCLAUD, C., LEDUC, J., d'HERBES, J. and THOMERE. R., "Injection de gaz miscible à haute pression à Hassi-Messaoud". *World Petroleum Congress,* Mexico, 2-9 April 1967.

2 DELCLAUD, C. and LEDUC, J., "Premiers résultats de l'injection de gaz à Hassi-Messaoud". Compte rendus du deuxième colloque ARTFP, 31 mai-4 juin 1965. Editions Technip, Paris 1965.

3 "L'injection de gaz à Hassi-Messaoud". *Rev. Franç. de l'Energie,* janvier 1965.

4 Studies by *Geopetrole* for *SN Repal* (unpublished).

5 Studies by *Francorelab* for *SN Repal* (unpublished).

6 COOK, A.B., WALKER, C.J. and SPENCER, G.B., "Realistic *K* values of C_7+ Hydrocarbons for calculating oil vaporization during gas cycling at high pressures." *JPT.*, July 1969, p. 901-915.

7 ATTRA, H.D., "Non equilibrium gas displacement calculations". *JPT*, Sept. 1961, p. 130-136.

8 JACOBY, R.H. and BERRY, V.J. Jr., "A method for predicting pressure maintenance performance for reservoirs producing volatile crude oil". *Trans. AIME,* 1958, 213, p 59-64.

4

miscible drive

41. INTRODUCTION

Under the heading of miscible drive come those fluid displacements during which the displacing fluid and the displaced fluid become miscible in all proportions, at least to a local extent.

We have seen that the overall displacement efficiency may be expressed as:

$$E = E_s \times E_i \times E_d$$

The term E_d accounts for the trapping of oil by capillary forces in the pores invaded by the displacing fluid, and is known as the "microscopic displacement efficiency".

It is clear that the nature of the displacing fluid is of great importance in the optimization of oil recovery, and that a fluid miscible with oil is to be preferred. The interfacial forces will in this case be eliminated and the microscopic displacement efficiency E_d will be 100% in theory. Miscible displacement has been used for some time in the laboratory, where residual oil is removed from core samples by flushing with an appropriate solvent.

Unfortunately, the solvents currently used for this purpose in laboratory work are rather expensive (often more expensive than crude oil), and could hardly be used for continuous injection into a reservoir.

The practical interest of miscible displacement became apparent when it was discovered that:

(a) To attain miscibility it is sufficient to inject a "slug" of solvent of limited volume (a few percent of the swept pore volume) displaced by a much cheaper follow-up fluid.

(b) Under certain conditions of pressure, temperature and phase composition, various fluids may become miscible with reservoir oil.

We shall refer to the first type of miscibility as "absolute miscibility" and the second as "thermodynamic miscibility".

It should however be noted that miscible displacements are subject to the same limitations as immiscible displacements as far as areal and vertical sweep efficiency are concerned. In particular, if the displacing miscible fluid is a gas of high mobility, such phenomena as fingering, gravity segregation or channelling may occur, reducing or even eliminating the potential benefit of the process as compared with conventional methods such as, for example, water injection.

42. MISCIBLE SLUG FLOODING

In the type of miscible displacement where a solvent slug is first injected, a certain volume of solvent is placed in contact with the oil with which it is miscible, and is then followed up with a fluid C which is immiscible with the oil O but miscible with S (Fig. 42.1).

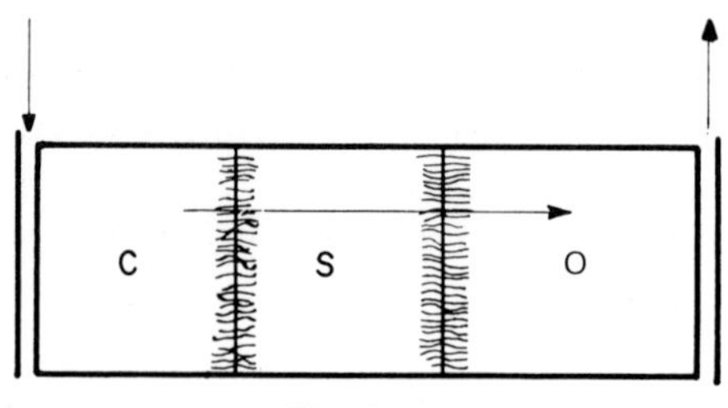

Fig. 42.1.

Typical displacement systems used are :

(a) Oil, LPG, gas.
(b) Oil, alcohol, water.

The theory of miscible displacement was explained in volume IV, where it was shown that between two miscible fluids in motion a mixing zone is formed, the size of which is proportional to the square root of time.

In the case of solvent slug injection, two mixing zones are formed, one between O and S, the other between C and S. These zones grow independently, each following a relation of the form indicated above. The size of the bank of pure solvent continually decreases as the sweep progresses. The volume of the solvent slug injected should be such that the bank of pure solvent is not exhausted before the O-S mixture breaks through at the production well, otherwise C and O, which are not miscible, would come into contact. When this happens it is known as a "miscibility rupture".

The major difficulty in planning a miscible displacement by solvent slug lies in the selection of an adequate slug volume: neither too small, to avoid the risk of a miscibility rupture, nor too large, to protect the economics of the project.

43. THERMODYNAMIC MISCIBILITY

The gas used to displace reservoir oil during secondary recovery by gas injection is almost always a mixture of hydrocarbons([1]). Thus the gas and oil both contain light hydrocarbons (principally methane), intermediate hydrocarbons (ethane to hexane) and heavy ends (heptane and above, or C_{7+}) but in completely different proportions.

Whereas an associated gas might contain (in mole percent) more than 80% methane and only 0.5% C_{7+}, even a very light oil, such as that from Hassi-Messaoud, will contain only 30 to 40% methane and more than 25% C_{7+} at reservoir conditions. On the other hand, if we consider the great variety of reservoir fluids encountered throughout the world we find two wide and practically overlapping ranges of composition for natural gases and crude oils. For example, a retrograde gas condensate high in condensibles and a very light oil have very similar compositions.

We can thus see that, during the injection of gas into an oil reservoir, as long as the original fluids are not completely different in composition, there will be a gradual exchange of components between the two fluids and their compositions will become more alike. Eventually part of the gas phase and part of the oil phase will no longer be separated by an interface and will thus become miscible.

In order for the exchange of components to be sufficiently great and lead to miscibility, a heavy oil will require an injected gas containing a high percentage of intermediate hydrocarbons, whereas a light oil will become miscible with a dry injected gas.

The phase exchanges are governed by the equilibrium constant K_i for each component *i*. K_i is defined as y_i/x_i and represents the ratio of the molecular fraction y_i of the component *i* in the vapour phase to the molecular fraction x_i of the same component in the liquid phase.

The equilibrium constant K_i is a function of pressure, temperature and convergence pressure P_K ([2]):

$$K_i = f(P, T, P_K)$$

For a given value of P_K, K_i tends to a value of 1 with decreasing temperature and with increasing pressure, at least for pressures not in the vicinity of P_K and at normal reservoir temperatures.

([1]) Sometimes inert or waste gases are injected, see later.

([2]) For further information see "Cours de Production", vol. II, Ecole Nationale Supérieure du Pétrole et des Moteurs (ENSPM).

A low value of K_i indicates that y_i is close to x_i, that is that the composition of the vapour phase is close to that of the liquid phase. Thus it is apparent that low temperature and high pressure are favourable conditions for the successful implementation of a miscible displacement project. For any given reservoir of course, the operator can influence only the pressure, and then only to a limited extent.

In the following Sections we shall describe in more detail the various types of miscible displacement using a graphical presentation of fluid composition known as a "ternary diagram".

44. THE TERNARY DIAGRAM

Thermodynamic miscibility can be more readily described if we represent complex mixtures of hydrocarbons by a combination of three arbitrary components made up of groups of hydrocarbons with similar thermodynamic properties:

(a) The light components, principally methane C_1 and possibly N_2, etc.

(b) The intermediate components, for example C_2-C_6 (all hydrocarbons from ethane to hexane included), and possibly CO_2 , H_2S etc. The intermediate hydrocarbons play a major role in thermodynamic equilibrium, all mixtures being described as rich or lean in these components.

(c) The heavy components, for example C_{7+} (C_7 and heavier hydrocarbons).

This division is completely arbitrary. We could for example take the groups C_2-C_4 and C_{5+} instead of C_2-C_6 and C_{7+}, and this would have little effect on the description of the phase behaviour.

Having chosen the three components, we can draw an equilateral triangle of which each apex represents one of the components.

Each side corresponds to zero % of the component represented by the opposite apex, and is divided into percentages of the component at one of its extremities (Fig. 44.1). The divisions of the three sides are made in a consistent fashion.

Let M be a point inside the triangle. This point represents a mixture of which, for example, the methane content is proportional to the distance between M and the side (C_{7+} , C_2-C_6). The methane content is read off the scale on the side (C_2-C_6 , C_1) by tracing a line from M parallel to the side (C_{7+} , C_2-C_6).

It can be shown that the point M representing all mixtures of two hydrocarbon systems M_1 and M_2 lies on the straight line M_1M_2 (Fig. 44.2).

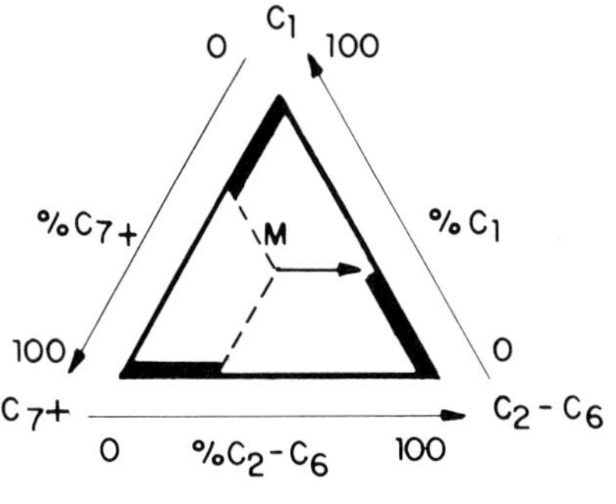

Fig. 44.1.

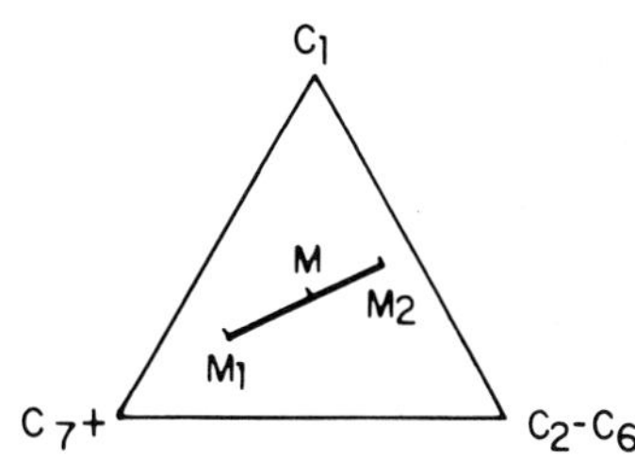

Fig. 44.2.

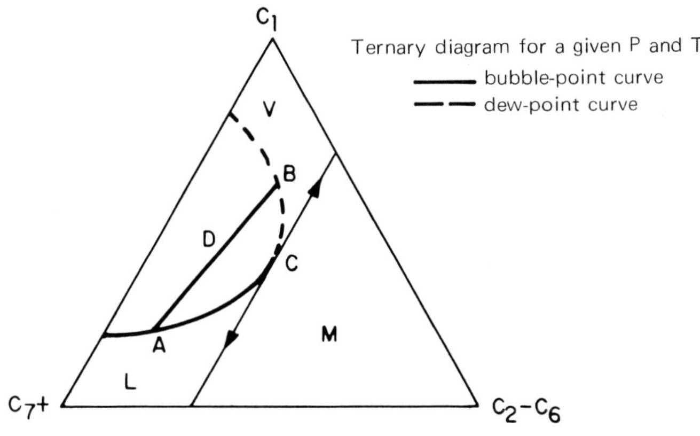

Fig. 44.3

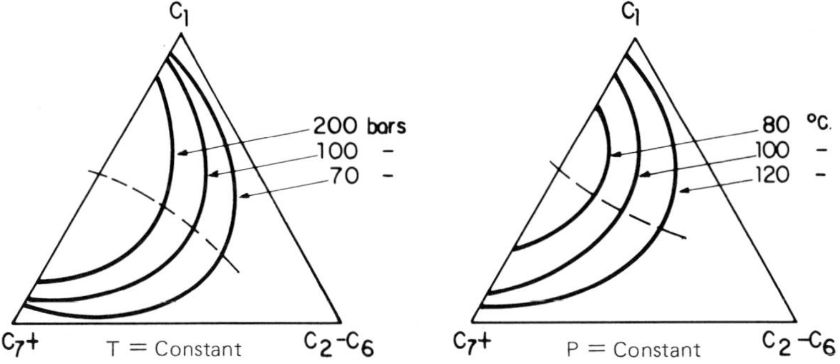

Fig. 44.4. Fig. 44.5.

At any given pressure and temperature, the point M may represent, according to its location inside the triangle, either a single phase or a diphasic fluid. We can trace on the ternary diagram, for a given combination of P and T, the curves bounding the diphasic region D: the bubble-point curve and the dew point curve. The composition corresponding to point C (Fig. 44.3) is that of the mixture of the three chosen components for which the given temperature and pressure are the critical temperature and pressure. For any given saturated liquid A there is a corresponding saturated vapour B with which it is in equilibrium. The line AB is known as a "tie line", and it is evident that each point on this line represents a disphasic mixture of which the compositions of the liquid and vapour phases are A and B respectively.

If we displace the tie line towards the critical point it tends to a tangent there. To the left of this tangent is the zone V in which the mixtures are in the vapour phase, and the zone L in which they are in the liquid phase. The hydrocarbon mixtures V and L are not miscible in all proportions. To the right of the tangent is the zone M in which the mixtures are termed "supercritical".

Mixtures in zone M are all mutually miscible and are also miscible with those of zones V and L so long as the line joining a point in zone M to a point in zones V or L does not intersect the dewpoint or bubble point curves.

Figure 44.4 shows how the size of the diphasic region varies with pressure at constant temperature.

Figure 44.5 shows how the size of the disphasic region varies with temperature at constant pressure.

It can be seen that high pressure and low temperature are very favourable conditions for miscible displacements, since they considerably reduce the size of the diphasic region. This has already been noted in Section 43.

45. BASIC METHODS OF MISCIBLE DRIVE

The main standard methods of miscible drive are:

(a) High pressure gas injection.
(b) Enriched gas injection.
(c) LPG slug injection.
(d) Alcohol slug injection.

45.1. High pressure gas injection

Two types of gas are commonly used in high pressure gas injection: natural (hydrocarbon) gases and inert gases.

45.11. High pressure natural gas injection

A. Phase conditions in the reservoir

Figure 45.111 illustrates the phase conditions during high pressure gas injection. The initial composition of the injected gas corresponds to point G.

Figure 45.112 shows the various stages in the formation of a miscible displacement front in the reservoir.

As will be shown, point O, corresponding to the oil composition, must lie to the right of the tangent at the critical point, **the oil must be rich in intermediate components.**

At the start of injection, the displacement is non-miscible and the line GO crosses the diphasic region. Thus some residual oil of composition O remains behind the gas-oil front.

The oil O and gas G are not in thermodynamic equilibrium. Phase exchange takes place and the result, at a given time and place, is gas of composition g_1 and oil of composition o_1. The gas becomes richer in intermediate and heavy components.

The oil O, while changing composition to that of o_1, tends to shrink. The oil saturation behind the front thus remains below the critical value and stays trapped, while the gas g_1 is displaced towards the front by the subsequent injection of gas G (State 2).

The gas g_1 comes into contact with newly formed residual oil of composition O. Since the fluids are not in equilibrium, phase exchange takes place and results in a gas g_2 and an oil o_2 which are in equilibrium, the gas g_2 being in contact with the front. The oil o_2, in contact with gas G, gives up more intermediates and its composition becomes o_a (State 3).

This progress continues until the composition of the gas in contact with the virgin oil becomes g_t, the point of contact of the tangent from O to the dew-point curve. At this stage miscibility between g_t and O has been achieved. From this point on, the displacement is miscible and no residual oil is left behind the front.

Behind the miscible bank the previously formed residual oils, of composition o_1, o_2 etc., continue to lose intermediates to the gas G. The limiting oil composition is o_p, on the tie line through G. The oil o_p can exchange no further components with gas G and is unrecoverable.

The experience of various operators indicates that a miscible bank is created after the injected gas has travelled a dozen metres or so from the injection well. The quantity of unrecoverable oil under these conditions will clearly be negligible.

High pressure gas injection is also known as "high pressure gas drive" and "vaporising gas drive".

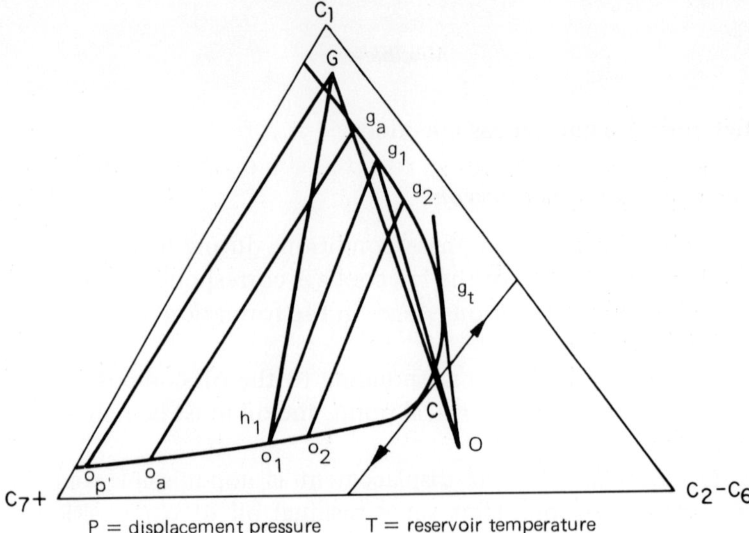

Fig. 45.111.

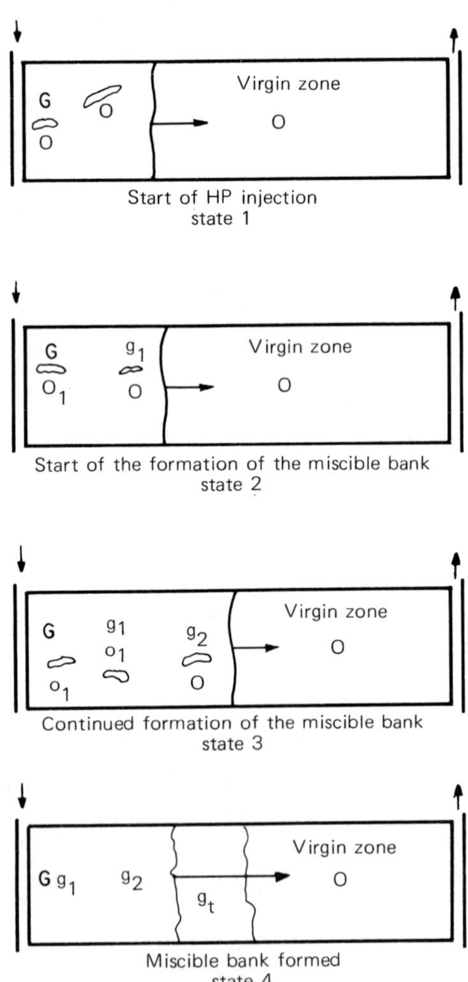

Fig. 45.112.

B. Miscibility pressure

On the ternary diagram drawn at reservoir temperature, miscibility can only be achieved between gas and oil of fixed compositions G and O respectively if the line Og_t is a tangent to the dew-point curve. This will only be the case at a pressure equal to or greater than the "miscibility pressure" P_m, at which the tangent at the critical point passes through O (Fig. 45.113).

The miscibility pressure for any particular combination of oil and gas is thus independent of the formation characteristics and displacement conditions. It can be determined experimentally using a high permeability artificial porous medium, in which high fluid velocities can be achieved and the experiments concluded in a reasonable time (Ref. 5). A long artificial porous medium (around 1-2 m, so that the mixing zone has time to form) is fixed vertically and saturated in oil. Using a sample of the gas to be injected, displacements are made at successively increasing pressures and the recovery achieved noted.

It will be found that recovery increases with pressure at first, then stabilises (Fig. 45.114). The pressure at which the slope of the curve changes is the miscibility pressure.

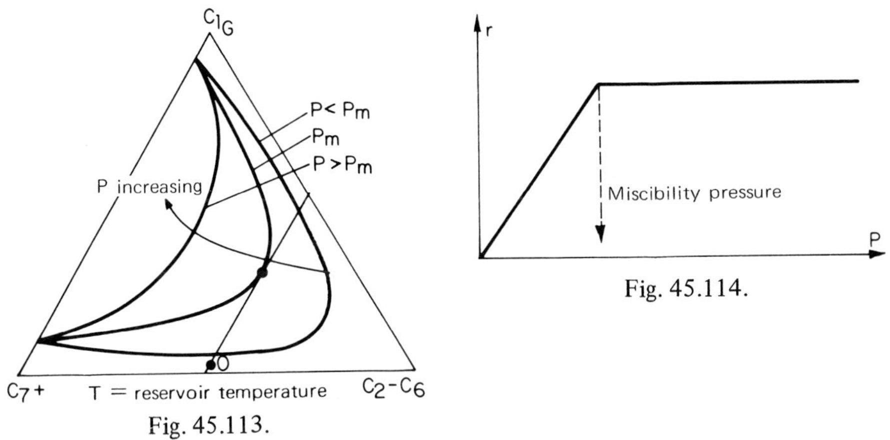

Fig. 45.113.

Fig. 45.114.

Each miscibility pressure determination, requiring 8 to 10 displacements, normally takes up to 5 days of laboratory work.

C. Application of high pressure natural gas injection

The miscibility bank is extremely stable. If a miscibility rupture occurs (due to heterogeneities, channeling, etc.) the miscibility bank reforms in the way described above.

The application of this process demands the following conditions:

(a) High reservoir pressure (deep formations). The minimum pressure required is of the order of 200-300 bar (3 000-4 500 psi).

(b) Oil rich in intermediates (gravity $\geqslant 35°$API).

45.12. High pressure inert gas injection

Once miscibility has been achieved, most of the gas injected in the course of recovery by miscible displacement is only needed to push forward the miscible front and fill up the porous medium.

It is thus possible to inject at first a limited volume of natural gas (around 5% of the pore volume) sufficient to ensure miscibility with the reservoir oil, and then to replace the injection of expensive natural gas with that of a cheaper gas, for example flue gas.

A suitable gas, approximately 12% CO_2 and 88% N_2, may be obtained by the combustion of relatively small volumes of separator gas. If we consider the combustion of methane for example:

$$CH_4 + 2\,O_2 + 8\,N_2 \rightarrow CO_2 + 2\,H_2O + 8\,N_2$$

9 volumes of combustion gas are obtained per volume of methane.

The inert gas also has the advantage of a high compressibility factor, thus the surface volumes of gas required are less than for natural gas. On the other hand, the inert gas must be treated and dried, which increases its cost.

It should be noted that it is possible to form a miscible bank by injecting from the start a mixture of natural and inert gas, or even an inert gas alone at higher pressure.

Figure 45.121 shows, for a given oil, the form of the variation of miscibility pressure for a methane-oil system when nitrogen is progressively substituted for methane.

The use of inert gas in miscible displacement also has application in two processes to be described: enriched gas injection and LPG injection.

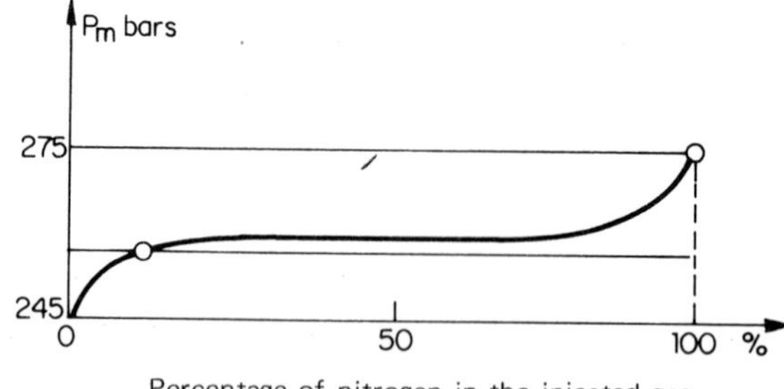

Fig. 45.121.

45.2. Enriched gas injection

A. Description of the process

In this case the formation of a miscible bank is achieved by way of the intermediate components in the natural gas. The process is also known as "condensing gas drive".

The injected gas is relatively rich in C_2-C_6, and is represented by the point G on the ternary diagram (Fig. 45.21); the oil in place is assumed to be heavy. (If the oil was light, i.e. rich in intermediates, a dry gas could be used.)

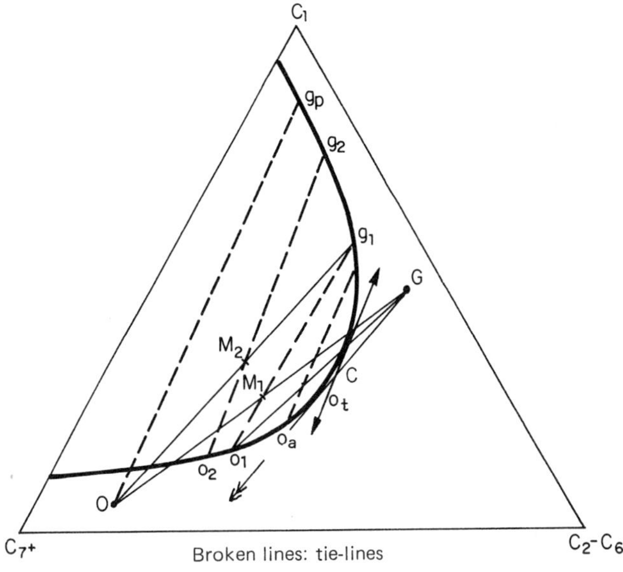

Fig. 45.21.

It can be seen that, as before, in order that miscibility be achieved at the operating temperature and pressure, the compositions of the oil O and the gas G must lie on opposite sides of the tangent at the critical point.

Figure 45.22 illustrates what happens in the reservoir during the displacement.

When enriched gas injection is started, the process is at first of the classical non-miscible type. Thus the residual oil O is in contact with the gas G (State 1).

By similar reasoning to that used in the case of high pressure injection, we can see that the oil behind the front will become progressively richer until it attains the composition o_t (Fig. 45.21), while the gas in contact with the virgin oil at the front becomes progressively drier than G (g_1, g_2, etc.) until by continual phase exchange with oil O it reaches the composition g_p (on the tie-line passing through O).

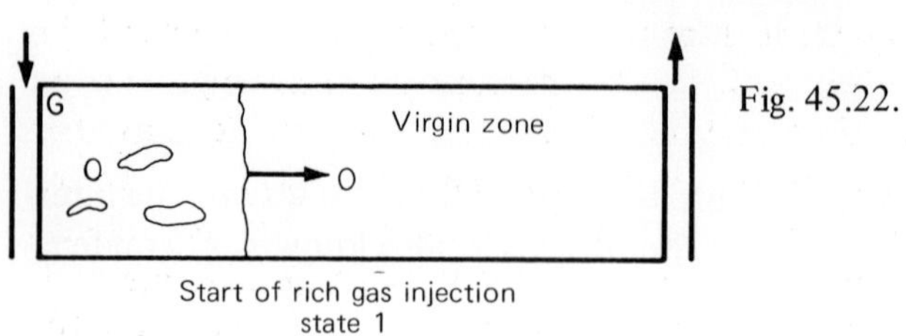

Fig. 45.22.

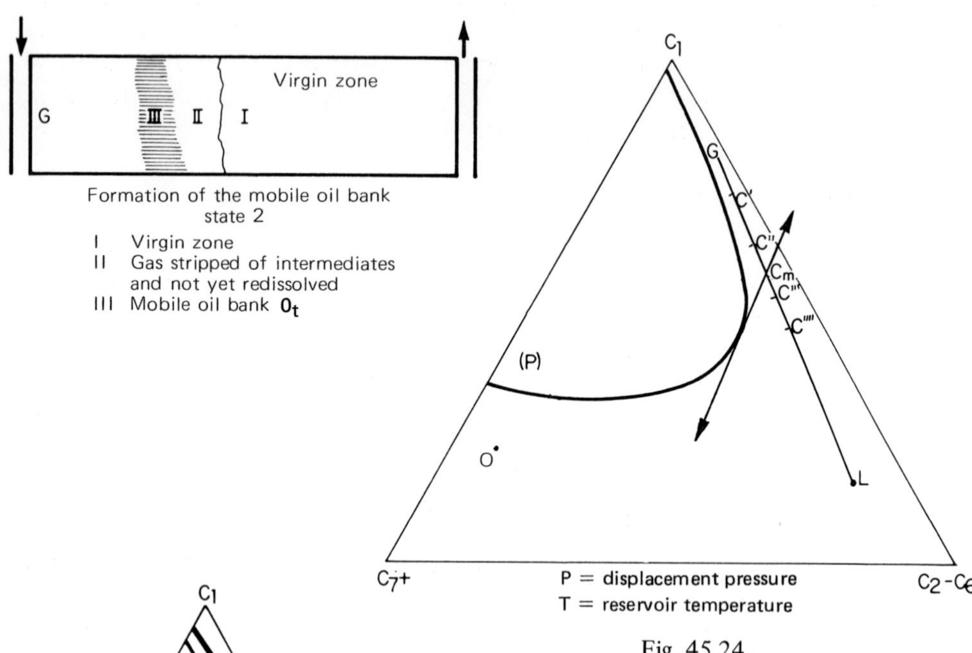

Fig. 45.24.

C_1
P < P_m
P_m
P > P_m
G_r
O
C
C
O
L
C_7+
T_G
C_2-C_6

Fig. 45.23.

As the composition of the oil changes from O to o_t the residual oil behind the front swells due to the absorption of light and intermediate components from the gas. At a certain stage the oil saturation will have increased sufficiently that the oil becomes mobile and a bank of oil of composition o_t will be formed.

At the end of this process there is no residual oil (assuming that asphaltenes and heavy components are not precipitated by the light alcanes), in contrast to high pressure gas drive in which the resulting heavy oil o_p is unrecoverable (Ref. 6).

The amount of dry gas ahead of the miscibility bank is continually reduced by absorption by fresh quantities of residual oil with which it comes into contact. If the oil is highly undersaturated the gas will be completely dissolved, if not some residual gas will remain stationary and will be bypassed by the miscibility bank, eventually being diluted in the driving gas.

B. Operating conditions

In the case of enriched gas injection the operating parameters are pressure and, possibly, the composition of the injected gas (which can be made richer by the addition of butane and propane or even LPG, the composition of which is indicated by the point L) (Figs. 45.23 and 45.24).

The rich gas may be obtained from a nearby reservoir or from one stage of the field crude separators, or may be a dry natural gas to which propane and butane have been added before injection.

1. Miscibility pressure for a given gas composition

On the ternary diagram drawn at reservoir temperature, miscibility can only be achieved between a rich gas G_r and an oil O of fixed compositions if the line $G_r o_t$ is a tangent to the bubble-point curve. This can only be achieved at a pressure equal to or greater than the miscibility pressure P_m, at which the tangent at the critical point passes through G_r.

The miscibility pressure for the gas of composition G_r is determined experimentally, as in the case of high pressure gas injection.

2. Composition C_m required for miscibility at a given pressure

If the gas is composed of a mixture of G and L, the first composition C_m at which the fluids are miscible at pressure P is as shown on Fig. 45.24.

In practice, miscibility pressures are determined for a range of compositions C', C'', C''', C'''' and a plot of composition vs. miscibility pressure drawn. The required composition C_m can then be read from the curve at the pressure P.

From a ternary diagram it can readily be appreciated that the richer the gas used, the lower the operating pressure may be, all other things being equal.

However propane and butane are expensive, and their continuous injection could hardly be considered. As soon as a miscible bank has formed, the injection of rich gas is replaced by the injection of dry gas miscible in all proportions with the rich gas behind the front.

It should be noted that the miscible bank does not have the same stability as in high pressure injection, due to the way in which it is formed. In the case of high pressure injection, the components required to form the miscible bank exist in every part of the reservoir, since they are contained in the oil in place. However, in enriched gas injection, the necessary components are not found in the reservoir but are obtained from the injected rich gas.

Thus the injection of rich gas should not be stopped until there is a sufficient reserve of rich gas behind the miscibility front. The volume of rich gas required depends on the nature of both the gas and oil involved. The further apart the compositions of the gas and oil on the ternary diagram, the greater the volume of rich gas required to form the miscible bank.

Enriched gas injection is most suited for the displacement of oil containing only small quantities of intermediates, when the reservoir temperature and pressure are moderately high. Displacement pressures usually fall between 140 and 210 bar (2 000-3 000 psi).

45.3. LPG slug injection

In this method, the miscible bank is formed at the outset by the injection of LPG of composition *L*, followed by the injection of dry gas *G* (Fig. 45.31).

The LPG is fully miscible with the reservoir oil in place *O*. It will also be miscible with the driving gas as long as the reservoir pressure is higher than the critical pressure of the gas-LPG mixture (Fig. 45.32), that is as long as the line *LG* does not intersect the dew-point curve.

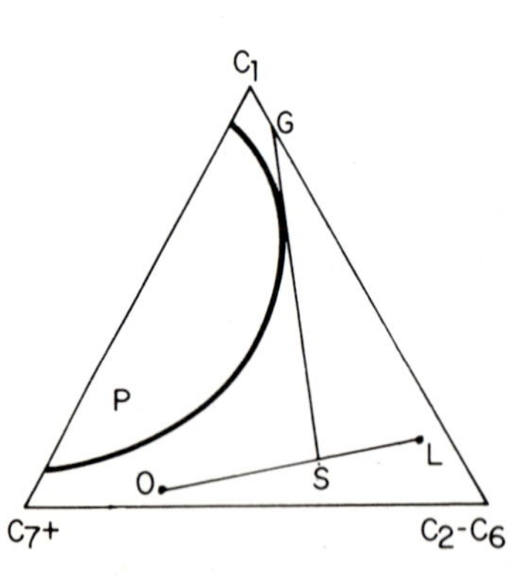

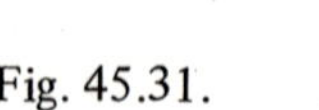
Fig. 45.31.

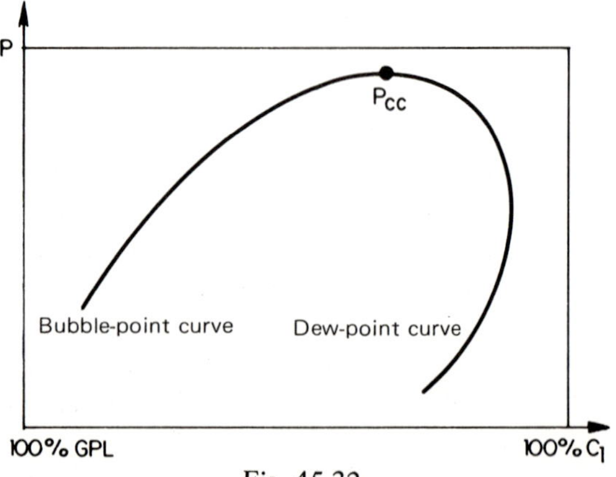

Fig. 45.32.

The LPG slug picks up reservoir oil during the course of the displacement and its composition changes along the line *LO*.

The LPG/oil slug remains fully miscible with the gas as long as its composition does not pass point *S*. The LPG content of the slug may be reduced to 30 or 40% and still remain miscible with methane. In many cases the LPG slug has been reduced to as little as 2% of the displaceable pore volume without reducing the recovery efficiency. In practice slugs of 2-9% of pore volume are used.

According to W.F. Kieschnik Jr., the minimum pore volume of LPG required is proportional to the square root of the distance to be covered, so that the larger the distance the smaller the relative volume of LPG needed.

45.4. Alcohol slug injection

Most miscible displacement processes, such as those we have already discussed, suffer from the disadvantage that high reservoir pressures are required, at least of the order of 100 bar (1 500 psi). Thus these methods cannot be used in shallow reservoirs, in which the pressures are low and which cannot always be recompressed because of the risk of fracturing the formation. Besides, the areal sweep efficiency is relatively poor because of the large mobility contrasts between gas, solvent and oil. In addition, natural gas and LPG are not always available in sufficient quantity in the vicinity of an oil field.

These constraints have led to the search for methods of miscible displacement in which water is the driving fluid. An obvious possibility is the use of alcohols as a slug between the oil and the water, since they are miscible with both liquids.

However, the principal difference between this method and those previously discussed is that the interstitial water would be displaced by the alcohol, whereas in miscible displacement by gas injection the interstitial water is unaffected. Thus the alcohol slug would be progressively diluted and, below a certain critical alcohol concentration, would no longer be miscible with the oil, at which stage the displacement would simply be water injection.

Several alcohols and combinations of alcohols have been the subject of laboratory studies, principally at Pennsylvania State University. At first isopropyl alcohol was studied. This has the disadvantages of being expensive and of absorbing water very rapidly, thus reducing its efficiency. Around 13% of the displaceable pore volume is required to ensure almost total recovery of the oil.

Other studies have shown that part of the isopropyl alcohol can be replaced, at the leading and trailing edges of the slug, by methyl alcohol. The methyl alcohol rapidly absorbs water, leaving the isopropyl alcohol at the centre of the slug practically water-free and thus retaining its oil displacement efficiency. A slug made up of three equal parts, each being 4% of the displaceable pore volume, with the central part of isopropyl alcohol and the outer parts of methyl alcohol, has the same efficiency as a 13% slug of pure isopropyl alcohol. As methyl alcohol is much cheaper than isopropyl alcohol, this combination is closer to being a commercial proposition.

Finally, if normal butyl alcohol is used in front of and methyl alcohol behind the isopropyl alcohol, the total slug volume required is reduced to 10% pore volume. However, the cost of butyl alcohol is prohibitive.

Even though this type of miscible displacement has not yet found commercial application due to the high cost of the various alcohols studied, the advantages of the method are evident, and the discovery of economically attractive processes should still be regarded as possible.

46. IMPROVED MISCIBLE DRIVE METHODS

It has been shown that the injection of natural gas under conditions leading to miscible displacement suffers from the following disadvantages:

(a) Poor vertical sweep efficiency E_i in heterogeneous formations (mainly due to the rock not being wetted by the gas).

(b) Poor areal sweep efficiency E_s (due to an unfavourable mobility ratio).

To improve matters, the following two methods may be used:

(a) Pre-injection of water.

(b) Chasing the miscible slugs with water.

46.1. Pre-injection of water (Ref. 3)

The injection of a solvent (enriched gas, LPG) in a stratified reservoir normally results in the most permeable layers receiving many times the solvent volume required to achieve miscible displacement throughout the field, before the least permeable layers have even received the minimum volume required. The distribution of solvent in the formation governs the fraction of the reservoir which may be miscibly swept. By reducing the effective permeability contrast the overall miscible displacement efficiency can be improved. This can be achieved by the pre-injection of water.

During the pre-injection of water the most permeable zones take more water than the least permeable zones. If there is a favourable mobility ratio ($M < 1$), the injectivity to solvent in the most permeable zones suffers a greater reduction than that in the zones of lower permeability. The result is a more even distribution of the solvent subsequently injected. Since the major influence on injectivity is the zone immediately surrounding an injection well, only a relatively small volume of water is required. Figure 46.11 shows the improvement in sweep efficiency calculated for the Lobstich Cardium unit reservoir of the Pembina field.

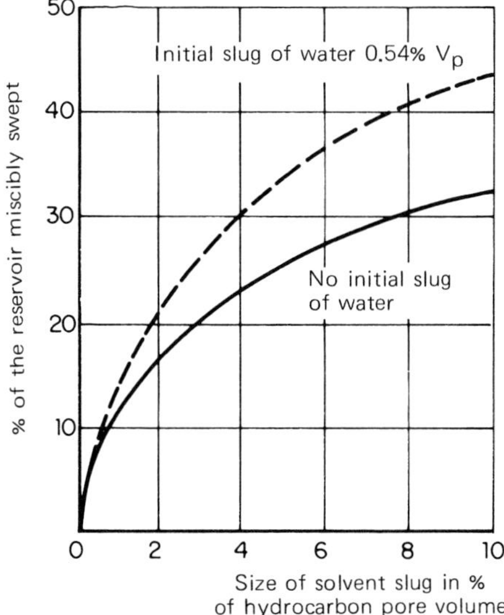

Fig. 46.11.

46.2. Miscible slugs driven by water

In miscible displacement by gas the gas-oil mobility ratio is often very unfarourable (it may be greater than 10) and thus the sweep efficiencies E_s and E_i are poor.

The mobility ratio may be reduced by injecting water with the gas, either simultaneously or alternately. Studies performed many years ago indicated that the lower the gas-water ratio, the lower the mobility ratio obtained. The lower limit for the gas-water ratio is that at which gas and water have equal velocity in the reservoir. If gas-water ratio lower than this limit is used, water will by-pass the injected gas, come into contact with the solvent slug, and the miscible displacement will revert to water injection, with a correspondingly reduced microscopic displacement efficiency E_d.

The required gas-water ratio can be estimated with the aid of a fractional water flow curve for the particular porous medium (Fig. 46.21).

If u is the total filtration velocity, the gas-oil mixture advances with a velocity given by:

$$W_m = \frac{u}{\phi(1 - S_{wm})}$$

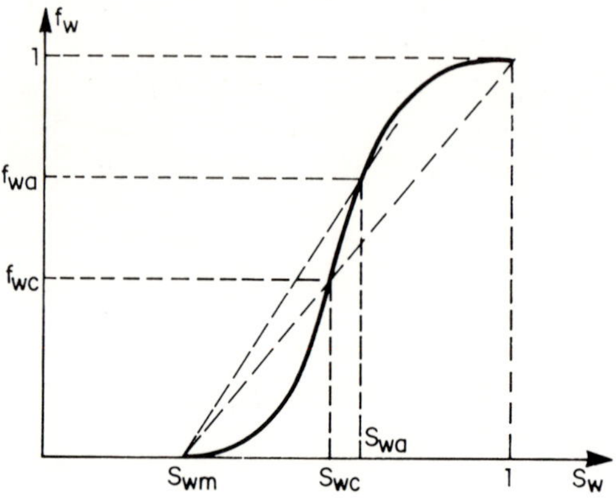

Fig. 46.21.

If S_{wa} is the water saturation near the injection wells, taking account of the ratio of the gas and water injection rates, the stabilised capillary zone advances at a velocity given by:

$$W = \frac{u f_{wa}}{\phi(S_{wa} - S_{wm})}$$

Let S_{wc} and f_{wc} be the co-ordinates of the point of intersection of the line joining the points $(S_{wm}, 0)$, $(1,1)$ with the fractional flow curve. It is apparent that:

(a) if $f_{wa} > f_{wc}$, $W > W_m$: the gas-water capillary zone will tend to catch up with the gas-oil mixture and the water will come into direct contact with the oil.

(b) if $f_{wa} < f_{wc}$, $W < W_m$: the gas-water capillary zone advances at a lower velocity than the gas-oil mixture. In this case the water will not come into contact with the oil.

47. BENHAM'S CORRELATIONS

Benham studied the maximum percentage of methane that could be used in enriched gas injection without losing miscibility (Ref. 8).

He established a correlation between methane percentage and temperature, pressure, molecular weight of C_2+ in the injected fluid, and molecular weight of C_5+ in the reservoir fluid (Figs. 47.1-47.12).

Consider a reservoir at a temperature of 104° C (220° F) containing an oil with a C_5+ molecular weight of 200.

If the available rich gas has a C_2+ molecular weight of 44 (e.g. propane) and the desired reservoir pressure is 175 bar, a miscible displacement could be achieved, according to Fig. 47.6, using a mixture containing rich gas and up to 52 % methane.

48. MODEL STUDIES

Miscible displacements have been studied by means of both physical and numerical models.

48.1. Physical models (Refs. 6 and 7)

A typical physical model consists of an arrangement of full diameter core samples saturated with reservoir fluids. Displacements are performed at fluid velocities identical to those to be obtained in the reservoir, respecting the physiochemical and thermodynamic exchanges between the various phases present.

The experiments are long and expensive, sometimes taking up to a year to perform.

48.2. Mathematical models

There are numerical compositional models in use today which take account of diffusion, convection and mass transfer due to thermodynamic exchange between phases. These models may be used to study miscible displacement projects.

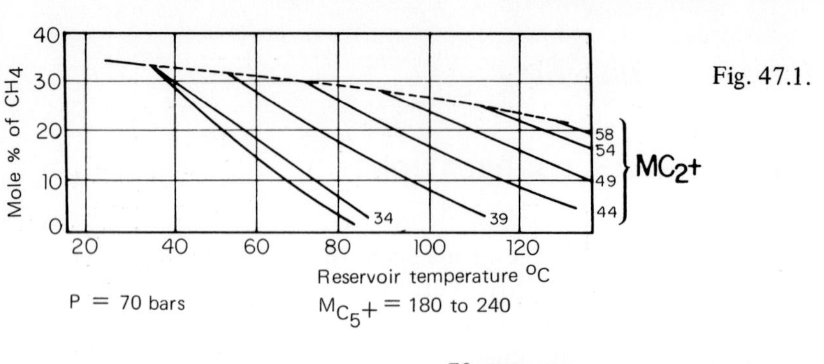

Fig. 47.1.

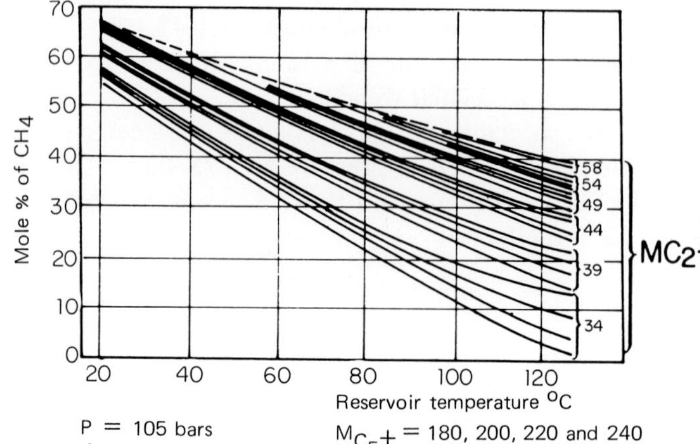

Fig. 47.2.

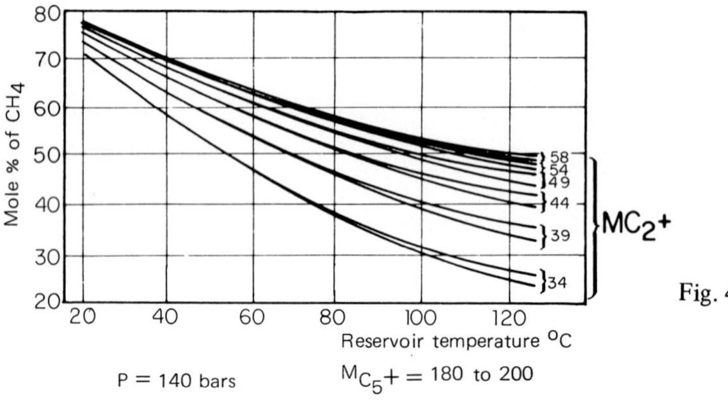

Fig. 47.3.

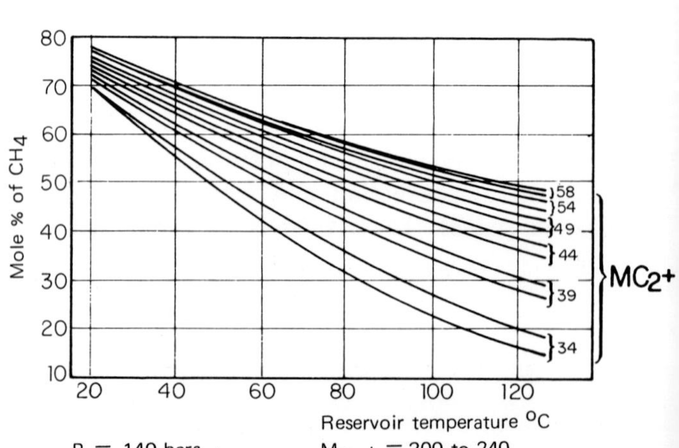

Fig. 47.4.

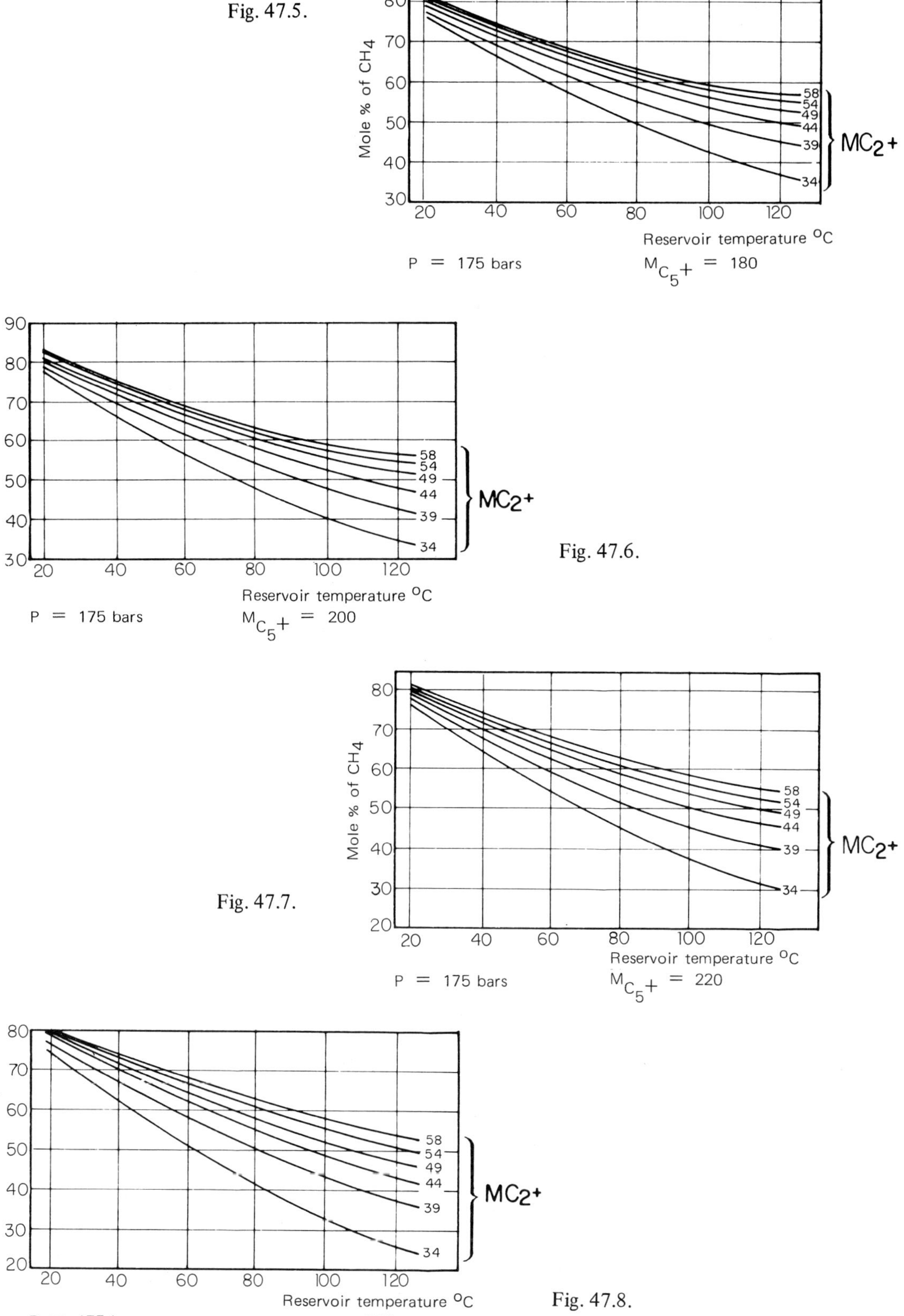

Fig. 47.5.

Fig. 47.6.

Fig. 47.7.

Fig. 47.8.

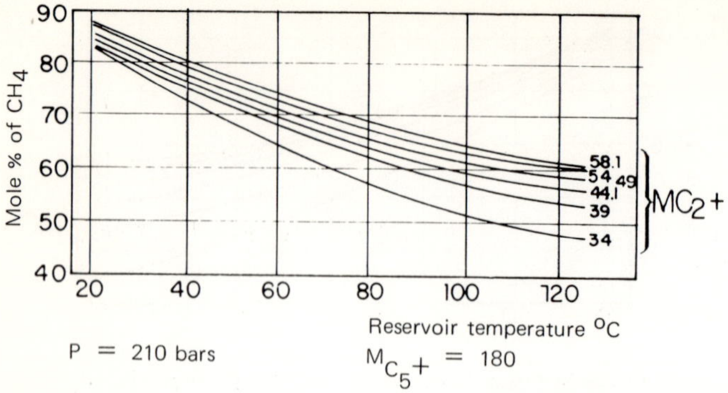

Mole % of CH_4
90
80
70
60
50
40
20
40
60
80
100
120
58.1
54
49
44.1
39
34
MC_2+
Reservoir temperature °C
P = 210 bars
M_{C_5+} = 180

Fig. 47.9.

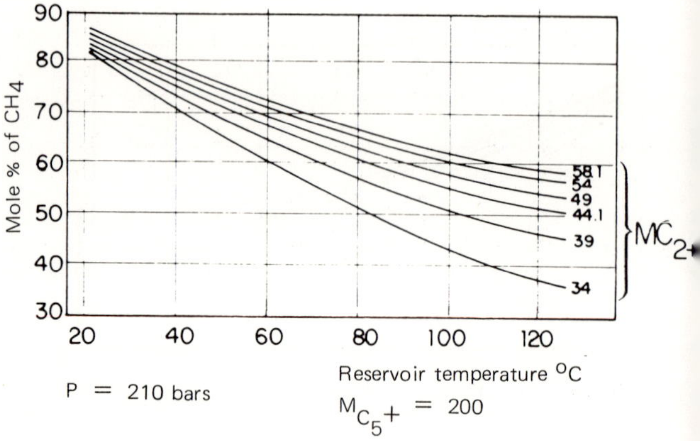

Mole % of CH_4
90
80
70
60
50
40
30
20
40
60
80
100
120
58.1
54
49
44.1
39
34
MC_2+
Reservoir temperature °C
P = 210 bars
M_{C_5+} = 200

Fig. 47.10.

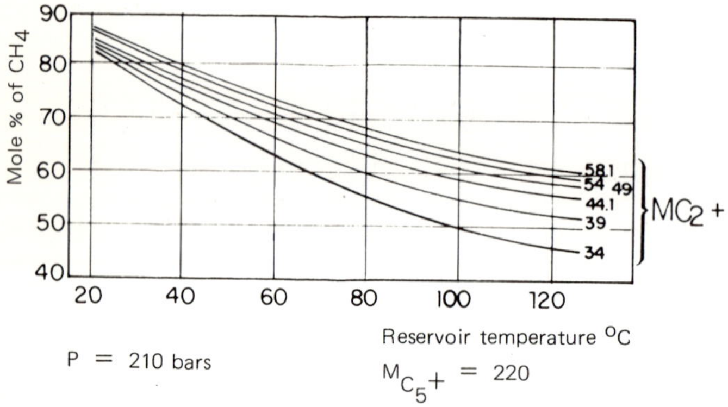

Mole % of CH_4
90
80
70
60
50
40
20
40
60
80
100
120
58.1
54
49
44.1
39
34
MC_2+
Reservoir temperature °C
P = 210 bars
M_{C_5+} = 220

Fig. 47.11.

Mole % of CH_4
90
80
70
60
50
40
30
20
40
60
80
100
120
58.1
54
49
44.1
39
34
MC_2+
Reservoir temperature °C
P = 210 bars
M_{C_5+} = 240

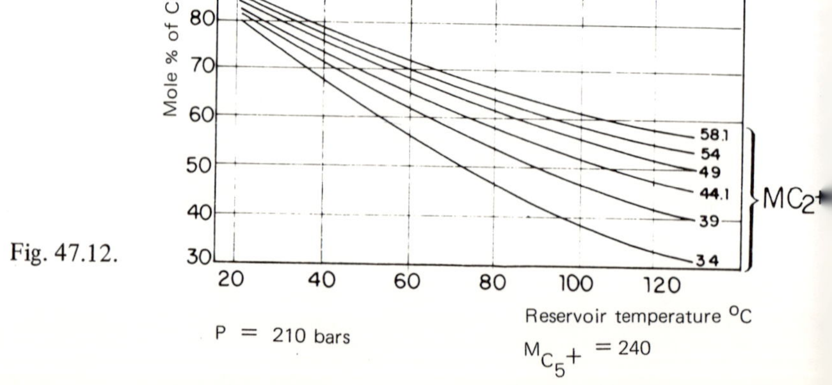

Fig. 47.12.

REFERENCES

1 MANASTERSKI, G., "Les déplacements miscibles ou miscibles drives". *Rev. Inst. Franç. du Pétrole,* mars 1960, p. 480-494.

2 CRAWFORD, P.B., "A review of many improved recovery methods". SPE . Paper 2 849, 1970.

3 CRAIG Jr., F.F., "A current appraisal of field miscible slug projects". SPE . Paper 2 418, 1969.

4 POTTIER, J., DELCLAUD, C., LEDUC, J., d'HERBES, J. and THOMERE, R., "Injection de gaz miscible à haute pression à Hassi-Messaoud". *7th World Petroleum Congress,* Mexico, 1967. Proceedings. Elsevier Publishing Co.

5 DEFRENNE, P., MARLE, C., PACSIRSZKY, J. and JEANTET, M., "The determination of pressures of miscibility". SPE Paper 116, 1961.

6 ALONSO, M., MORINEAU, Y. and BEACOM, G., "Rôle de l'expérimentation en conditions de gisement dans l'étude du comportement physique des réservoirs et les prévisions d'exploitation". Communication n° 21. Premier Colloque ARTEP, juin 1971.

7 SIMANDOUX, P., PACSIRSZKY, J., "Rainbow field. A general survey about the experimental study made in IFP. "Rapports IFP 18 062-I et 18 062-II, avril 1970.

8 BENHAM, A.L., DOWDEN, W.E. and KUNZMAN, W.J., "Miscible fluid displacement. Prediction of miscibility". *Trans. AIME,* 219, 1960, p. 229-237.

9 LARSON, V.C., PETERSON, R.B. and LACEY, J.W., "Technology's Role in Alberta's Golden Spike miscible project". *7th World Petroleum Congress,* Mexico, 1967. Proceedings. Elsevier Publishing Co.

10 THIBIERGE, Ph., "Nouvelle mise au point sur l'opération d'injection de gaz à Hassi-Messaoud". Comptes rendus du 3e Colloque ARTFP, Pau, 1968, Editions Technip.

11 GIRAUD, A. and THOMERE, R., "Injection de gaz à haute pression. Premiers résultats sur l'efficacité de balayage en five-spot". Comptes rendus du 3e Colloque ARTFP, Pau 1968, Editions Technip, Paris 1969.

REFERENCES

5

gas recycling in gas-condensate reservoirs

51. INTRODUCTION

A gas-condensate reservoir may be developed in one of two ways:

(a) By producing the reservoir by natural depletion. The produced fluids are processed and the resulting dry gas and gasoline sold.

(b) By reinjection of all or part of the dry gas obtained back into the reservoir. The attractiveness of this recycling technique depends on the particular circumstances, and on how the recycling is carried out.

Development by natural depletion leads to the formation of an increasing liquid saturation in the pores, until the start of revapourisation which is in any case only partial. The percentage of heavy components in the production decreases from the time that the dew point is reached until the retrograde pressure is attained. At this point the percentage increases but only very gradually. Thus the gasoline production is a decreasing function of time (Ref. 1).

These phenomena, very apparent in limited reservoirs, tend to become less distinct in reservoirs with active aquifers. However, as was noted in Volume V of this series (1), gas trapped at high pressure below a rising oil-water contact represents a significant loss, both of dry gas and condensate. For example, the mathematical model simulation of the natural depletion of unit 1 of the South Kaybob field in Canada, which has an active aquifer, indicated a gasoline recovery factor of the order of only 60 % (Refs. 4, 5 and Appendix 5.).

---

(1) HOUPEURT A., *Estimation des réserves récupérables par drainage naturel.* Cours Ecole Nationale Supérieure du Pétrole et des Moteurs (ENSPM), tome V, Editions Technip, Paris 1968, page 188.

Re-injection into the formation of dry gas obtained after processing the produced fluids avoids the disadvantages of natural depletion. The reservoir pressure is partially maintained, loss by retrograde condensation is reduced and the composition of the produced fluids remains fairly constant. The dry gas injected may cause hydrocarbon liquids already condensed in the reservoir to re-evaporate, but on the other hand the gas may break through prematurely at the production wells, and in this case the recycling will only benefit part of the reservoir. If an aquifer is present, recycling will retard the rise of the oil-water contact, thus reducing the loss of trapped hydrocarbons.

Even if all the available dry gas is recycled, the resulting pressure maintenance will only be partial, since the volume of gas produced is greater than that re-injected. The difference is made up of gasoline and the gas used to power the compressors. For example, in unit 1 of the South Kaybob field, re-injection of all the dry gas produced only replaces 74% of the reservoir voidage.

Full pressure maintenance can sometimes be realised by the use of make-up gas from a nearby source (dry gas reservoir or associated gas from an oil reservoir).

For a gas-condensate reservoir with an active water drive, partial replacement of the reservoir voidage is often the optimum method of development. Too small a replacement increases both the amount of gas trapped by the encroaching water and the losses due to condensation. On the other hand a high rate of replacement is uneconomic: it delays the dry gas sales and increases the cost of compression. For unit 1 of the South Kaybob field a study of the effect of the rate of replacement on gasoline recovery was made and the following results obtained:

Percentage replacement	Gasoline recovery (%)
100	84
74 (full recycling)	82
60	80.7

Note that the recovery factors are much higher than for natural depletion.

One further parameter may be controlled by the operator when developing a field by gas recycling: the point at which gas re-injection commences. Recycling may be initiated from the start and thus at high pressure (frequently higher than or equal to the dew-point pressure), or at a lower pressure following a period of natural depletion. The best solution can only be determined by a study of both the technical and economic aspects of the particular development.

A gas recycling project generally ends with a depletion phase after the dry gas has broken through at the production wells. This allows the recovery of the condensate from the unswept zones. The depletion is continued until the abandonment pressure is reached. Typically, this pressure might be around 2 bar/100 m depth (10 psi/100 ft), although lower pressures could be reached if compressors were used to pump the gas to the treatment plant.

52. THE THERMODYNAMICS OF GAS RECYCLING

As we have seen, gas recycling may take place either:

(a) At a pressure P higher than or equal to the dew-point pressure P_d. In this case there are no thermodynamic effects, the gas is displaced at constant composition (original composition), or

(b) At a pressure lower than the dew-point pressure and possibly decreasing with time for technical or economic reasons (insufficient quantities of available gas or a desired reduction in the cost of compression).

In the second case the problem is to determine the composition profile (liquid saturation and gas richness) between the injection and production wells.

Standing *et al.* (Ref. 2) have suggested a method of calculation which assumes complete displacement ($E_d = 1$) of saturated gas by dry gas in each zone swept and which neglects volume changes due to the pressure gradient between injection and production wells.

The region between an injection well and a production well is first divided into N cells of equal volume.

Let us consider a time at the start of recycling at which the first cell adjoining the injection well has just been invaded by dry gas. The gas will be in contact with the hydrocarbon liquids previously condensed there (since $P < P_d$) and will thus become enriched with heavy components. If the equilibrium ratios are known an equilibrium calculation may be performed. All the enriched gas then passes into the second cell where it again comes into contact with condensed liquids. A new equilibrium calculation may be made and so on. These calculations may be undertaken by computer, since it is normal to use about a dozen cells.

A composition profile is thus obtained and, if the assumption is made that the area drained by the dry gas between the injection and production wells is constant (the required assumption for linear displacement), the variation of the composition of the produced fluid with time may be estimated.

The results obtained by Standing are shown on Fig. 52.1.

In reality, the above calculations are generally a rather rough approximation. The area drained varies continuously and part of the reservoir unswept at the time of breakthrough is subsequently swept by the dry gas, thus the percentage

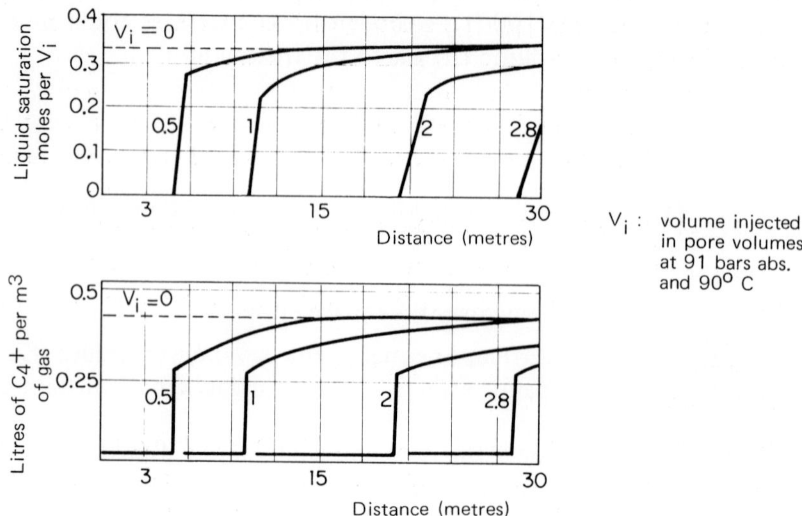

Fig. 52.1. An example of the composition profile for linear flow (From M.B. Standing).

of heavy components in the production is usually higher than that calculated by Standing's method.

53. SWEEP EFFICIENCY

The displacement of wet gas by dry gas is essentially a miscible displacement at a mobility ratio of 1. Although the two gases will obviously have different compositions they will both have methane as their principal component and the effect of the viscosity ratio can in practice be neglected.

Thus the displacement efficiency E_d may be assumed to be 100%, especially if the recycling takes place at or above the dewpoint pressure. If $P < P_d$ the displacement efficiency E_d will be less than 1 due to the presence of the stationary liquid phase and its incomplete re-evaporation by the injected dry gas.

The invasion (vertical sweep) efficiency E_i is dependent on the reservoir characteristics but is usually good. 70-80% or better. However, there may be a significant difference between the specific gravities of the dry and rich gases, and the dry gas may channel preferentially along the top of the bed.

The overall efficiency at breakthrough is thus of the order of 60% :

$$E = E_s \times E_i \times E_d = 0.7 \times 0.9 \times 1 = 0.63$$

If injection is continued after breakthrough it is possible to sweep practically the entire reservoir, but the gasoline fraction will continually decrease and the operation will finally become uneconomic.

Models can be used to obtain a better estimate of E_s for a particular field. Originally only analogue models were available, but with the advent of the modern computer these have now been almost totally replaced by mathematical models.

The following assumptions were necessary in analogue model studies:

(a) A vertical interface exists between displacing and displaced phases.

(b) Where two fronts exist (e.g. gas/gas and gas/water) one is considered to be stationary.

(c) The average reservoir pressure is constant.

(d) Gravity may be neglected.

The first assumption is still made today in many mathematical models (see example in Appendix). The other assumptions need not be made in mathematical models and this is one of the advantages they have over analogue models. However, in practice the third assumption is not often very important, since during recycling reservoir pressures and flowrates are held constant for long periods of time.

54. WELL LOCATIONS

There are two cases to be considered when studying the location of new wells and the use of existing wells as injectors or producers, depending on whether recycling starts after a long period of natural depletion or is planned from the discovery of the field.

In the first case the field is already developed, thus the physical characteristics and the geometry of the field are reasonably well known. From a numerical model representing the reservoir the gas sweep efficiency can be obtained as a function of the use of the existing wells as injectors or producers, taking account of the different capacities of the wells. The maximum sweep efficiency may then be obtained by restricting the rate of certain injection or production wells, and possibly by adding some extra wells.

In the second case an additional parameter is available for use in the model study: the co-ordinates of the wells to be drilled. On the other hand, the reservoir limits are ill-defined and the well capacities unknown. As a first step, a model is constructed using the available geological maps, generally assuming that the reservoir has lateral homogeneity. The well capacities are taken to be proportional to the reservoir thickness, unit capacity being based on the results of production tests on the discovery well. The optimum arrangement of wells is then selected, being that which will give the highest recovery of wet gas for the lowest investment, i.e. for the smallest number of wells. As development of the field proceeds, more data become available. The model study may be repeated if necessary and the location of the remaining wells may possibly be adjusted.

There are some general guidelines which may be of use in such model studies:

(a) In long, thin structures, the injection wells are often located at one end of the field and the production wells at the other. In this case the displacement is practically linear and a high areal sweep efficiency is assured.

(b) In circular reservoirs displacement may take place radially (between central and peripheral rings of wells) or linearly (between three parallel lines of wells of which one line is central). The activity of the aquifer is generally unknown and, since the average reservoir pressure will decline during recycling(1), it is advisable to displace the gas towards the centre of the structure in order to avoid the production of significant quantities of water. The gas column is small on the periphery of the structure and wells completed there would have very low water-free production capacities. Injection wells do not suffer from this disadvantage and injection may take place into both gas and water-bearing zones without penalty.

(c) In the case of converging displacements, it is preferable to locate the injection wells and fix the individual well rates so as to obtain simultaneous breakthrough at the production wells. In this way the maximum quantity of heavy components is obtained most economically, since the treatment plant is working continuously at its maximum capacity.

The Kaybob reservoir is one example of a field in which the well locations were studied using mathematical models.

This reservoir, made up of two zones separated by a shale, is a long thin structure and has a large active aquifer.

Unit 1, one of the three concessions, was the first to be studied. Three injection well location schemes were examined:

(1) Injection into both zones, wells located at the extremities of the field.
(2) Selective injection, wells located at the extremities of the field.
(3) Injection into both zones, wells located at the centre of the field.

Six simulations were performed to cover the various well location and injection rate combinations, using a two-dimensional model.

---

(1) Unless a supply of make-up gas is available.

Central injection appeared to be the best solution. Most of the areas unswept at abandonment had relatively small gas columns. This scheme was the one selected, with the addition of an extra injection well at the southern end of the concession placed so as to avoid the loss of rich gas to unit 2. The alternative solution of injection at the extremities gave a poorer sweep efficiency and required a larger and therefore more expensive surface pipework system. More details are given in the appendix.

For the other two units, three injection schemes were studied using a three dimensional mathematical model which included the effects of gravity. These schemes were:

(1) In-line injection using three equally-spaced lines of wells perpendicular to the major axis of the reservoir, plus some additional injection wells at the southern end of the reservoir (Appendix 5, Fig. A.5.5 and A.5.6).

(2) Downdip injection (injection wells on the western flank) (Appendix 5, Figs. A.5.7 and A.5.8).

(3) Updip injection.

In this particular case, the justification for downdip injection was to be able to locate the production wells in the thickest, most productive areas, and the injection wells in the thinest, least productive areas. The updip injection scheme was intended to minimise the number of injection wells required and to take advantage of the effects of gravity: the re-injected gas, being lighter than the gas in place, would stay above it.

The results of the simulation indicated that the recovery from unit 2 would be highest with in-line injection, while unit 3 would be best developed by downdip injection. (Figs. A.5.3 and A.5.4).

55. PRODUCTION CONTROL

Theoretically, measurements of the liquid content([1]) of the production of each well as a function of time should be sufficient to determine the moment of breakthrough and the dilution of the wet gas by the injected dry gas. In practice however, many difficulties are encountered in the interpretation of the variation in liquid content:

- In the field only the separator gas/liquid ratio is normally determined. This ratio depends on the pressure and temperature of separation and the latter

---

([1]) In this section, liquid content refers to condensible hydrocarbon liquid, C_{5+} .

is not readily controlled by the operator, being a function of the ambient temperature and the flowrate. In order to obtain comparative figures it is thus necessary to take account of the liquid content of the separated gas (e.g. by passing the gas through a charcoal absorber).

• The production stream analysed may not in fact be representative of the reservoir fluid around the well-bore since the composition of the surface fluids will be affected by changes in flowrate. At high rates there will be a large pressure sink around the well-bore and consequently a significant amount of retrograde condensation, whereas at low rates there will be slippage between gas and liquid as the fluids travel up the production string. The surface fluid will not be representative of the reservoir fluid until a period of stabilisation at constant rate has elapsed (this period being between 24 hours and over a week, according to Lewis (Ref. 7)). This phenomenon can prove to be a problem, especially if the capacity of the test separator is less than the well's normal production rate.

• Finally, the gas sampled as development proceeds comes from further and further away from the well and it may be that the composition of the reservoir gas varies from point to point. For example, the Comstock sand reservoir in the Stratton field in Texas contained gas of which the liquid content varied from 106 to 40 litres/1 000 m^3.

These variations may be due to the way in which the hydrocarbons originally formed in the reservoir.

As a result of all this, in spite of the precautions taken during regular sampling (say every 3 months for each well), plots of the variation in liquid content with time are difficult to interpret.

There are also other factors which complicate the determination of breakthrough from measurements of liquid content:

(a) Firstly, the liquid content may decrease before breakthrough.

Figure 55.1 shows the variation of the average liquid content measured on 19 wells in the Gloria field, Texas, in which breakthrough had not occured. The decrease in liquid content with time is due to retrograde condensation in a unswept area of the reservoir, following the reduction in average reservoir pressure during recycling.

(b) Secondly, the decrease in liquid content after breakthrough (due to the wet gas production being diluted with dry gas) may be retarded and masked by the re-evaporation of some of the retrograde liquid earlier deposited in the reservoir.

However, breakthrough may be established more clearly using plots of condensate density and liquid content of the separator gas (assuming in the latter case that separation conditions are constant). These parameters remain practically constant until the arrival of the dry gas.

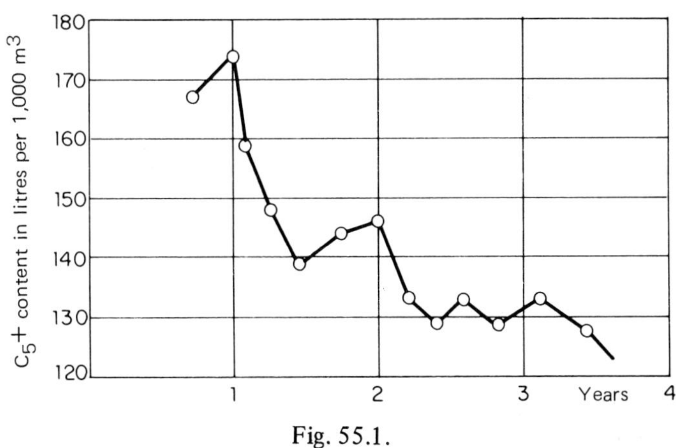

Fig. 55.1.

56. PRODUCTION EQUIPMENT

Production installations involve compression, separation and fractionation equipment.

Compression equipment has already been described in Chapter 3, Section 37.

By separation only part of the C_{6+} and a very small fraction of the C_5 present in the production stream can be recovered as a liquid. If more complete recovery of gasoline and LNG (propane and butane) is required, fractionation of the separator gas will be necessary, either by adsorption or at low temperature.

A rule of thumb used in the USA is that a temperature reduction of 1 °C in the gas stream increases the liquid recovery by 0.5 litres/1 000 m^3, for temperatures above − 7 °C. Below this temperature the increase in recovery is small.

Whereas it may be assumed that all the C_{5+} in the well stream will be recovered either as gasoline or condensate, the extraction of LNG from the separator gas will be determined by market conditions. For example, in the United States the extraction of propane was hardly considered until 1942. In comparison, as early as 1959 the recycling project at Northwest Branch, Louisiana, envisaged the recovery of practically all the condensibles.

It is interesting to note that Gulf Coast operators consider that it is uneconomic to treat gas with a condensible liquid content of less than 40 litres/ 1 000 m^3. This limit determines when a production well at which dry gas has broken through will be abandoned. For example, in a reservoir undergoing gas recycling above dew point, the wells will be abandoned when the fraction x of wet gas in the production is such that:

x (liquid content of wet gas) + $(1 - x)$ (liquid content of dry gas)

is equal to the economic limit of the treatment plant.

57. DETERMINATION OF OPERATING CONDITIONS

As far as the recovery of the C_{4+} is concerned the pressure at which recycling takes place (preceeded and followed by natural depletion phases) has practically no effect (due to the re-evaporation of the condensed liquids by the dry gas).

From an economic point of view everything depends on the market for the gas. The following points should be noted:

(a) For a given recycling rate, depletion is faster at low pressure since the total volume of dry gas required to re-evaporate the condensed liquids is lower. But as the production capacity of a well is given by:

$$Q = \alpha (P_e^2 - P_{wf}^2)^n$$

it will be necessary either to increase the number of wells or to reduce the bottom-hole flowing pressure P_{wf}. In the latter case the pressure losses in the surface pipework would have to be reduced, which would necessitate larger diameter and therefore more expensive flowlines.

(b) When recycling takes place at high pressure all the gas is sold at the end of the project, whereas when recycling at low pressure the gas is sold from the start.

In practice, all the different methods of development must be considered. Studies are therefore made of the cases: natural depletion from original pressure to abandonment pressure; recycling at the dew point followed by natural depletion; recycling at two or three intermediate pressures preceeded and followed by natural depletion. For each recycling pressure the optimum production rate must be determined.

For the different recycling pressures, the present worth of the production is determined together with the capital investment and operating costs. All solutions are compared and the most economically attractive solution is selected.

APPENDIX 5.1

An example of the use of mathematical models in the study of a recycling project in a gas-condensate reservoir South Kaybob Field (Canada)

I. RESERVOIR CHARACTERISTICS

The South Kaybob field was discovered in Canada in 1961. It is a thin structure about 51.5 km long bounded to the east by a fault (Fig. A.5.1).

The field covers an area of 230 km^2 and is divided into 3 units or concessions numbered 1 to 3 from North to South, which are not all operated by the same company.

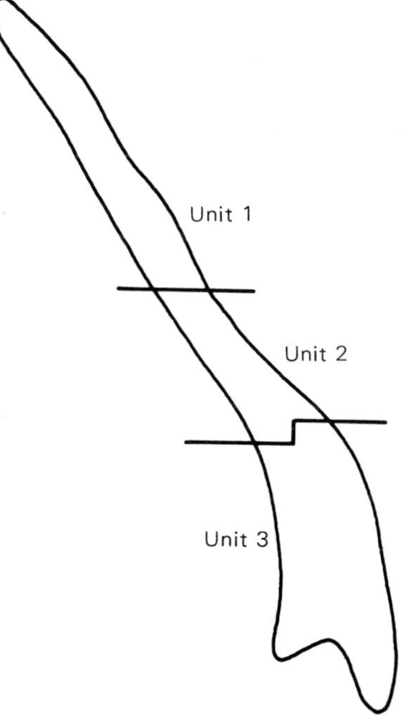

Fig. A.5.1.
(Ref. 5).

Throughout the field there are two reservoir zones separated by a shale of 1.5 m thickness. The formation dips to the south-west at an angle of 0.7 degrees. Other characteristics are as follows:

(a) Depth: 3 200 m.
(b) Initial reservoir pressure: 324 bar (4 700 psig).
(c) Reservoir temperature: 114 °C.
(d) Total thickness: 46 m.
(e) Large active aquifer.
(f) Gas in place: 113×10^9 m^3.
(g) Gas dew-point: 252 bar (3 655 psig).
(h) Gas richness: 650 litres condensibles/1 000 m^3.
(i) Sulphur content: 39 kg/1 000 m^3.
(j) Total gas gravity: 1.03.
(k) Total gas compressibility factor Z: 0.91 at 324 bar, 0.79 at 190 bar, 0.90 at 70 bar.
(l) Gravity of gas re-injected: 0.62 (free of H_2S and C_{5+}).

II. MATHEMATICAL MODEL STUDY OF UNIT 1

Although the field was discovered in 1961 it was developed slowly at first and it was not until 1967 that the simulation of unit 1 was undertaken.

A. The model

The model was constructed in order to predict the behaviour of a reservoir with several wells and an active aquifer undergoing natural depletion or recycling. It consists of two basic parts: the calculation of pressure distribution and the calculation of the areas swept.

1. Calculation of pressure distribution

Three steps are involved in this calculation:

(a) Solution of the diffusivity equation in two dimensions.
(b) Calculation of water influx.
(c) Calculation of limiting water-free production rates.

a. Solution of the diffusivity equation

The diffusivity equation for the flow of gas in a porous medium is:

$$\frac{\partial}{\partial x}\left(\frac{k_x h}{\mu Z}\frac{\partial p^2}{\partial x}\right) + \frac{\partial}{\partial y}\left(\frac{k_y h}{\mu Z}\frac{\partial p^2}{\partial y}\right) = \frac{\phi h c}{Z}\frac{\partial p^2}{\partial t} + 2q(x, y, t) \qquad \text{(Eq. A.5.1.)}$$

where

$$c = \frac{1}{p} - \frac{1}{Z}\frac{\partial Z}{\partial p} \qquad \text{(Eq. A.5.2.)}$$

Equation A.5.1. may be solved for a regular grid system using the alternating direction implicit method (ADIP).

The model is designed in such a way that the user may change flowrates and shut-in or add wells at each time step.

b. Calculation of water influx

The radial circular transient method of Van Everdingen and Hurst is used([1]).

The time steps are chosen in such a way that the pressure distributions corresponding to 10, 20, 30, 60, 90, 180, 270 . . . days are amongst those calculated. The cumulative volume of water influx is calculated at each time step.

The net water influx is distributed amongst the various cells according to their pressure and thickness:

$$(\Delta W_e)_c = \frac{(\Delta W_e)_N \left[1 - \left(\frac{Z_1}{Z_2}\right)\left(\frac{p_2}{p_1}\right)\right]_{i,j} h_{i,j}}{\sum_1^{N_x}\sum_1^{N_y}\left[1 - \left(\frac{Z_1}{Z_2}\right)\left(\frac{p_2}{p_1}\right)\right]_{i,j} h_{i,j}}$$

where

N_x is the number of cells in the x direction,

N_y is the number of cells in the y direction,

$(\Delta W_e)_c$ is the net water influx in cell i, j during a time step,

$(\Delta W_e)_N$ is the net water influx for the whole reservoir during the same time step,

p_1 and p_2 are the pressures at the beginning and end of the time step.

The water encroachment at the base of the reservoir is described by a thickness reduction technique:

$$(h)_{n+1} = (h)_n - \frac{(\Delta W_e)_c}{(\Delta x)^2 \phi (1 - S_w - S_{gr})}$$

([1]) HOUPEURT, A., *Estimation des réserves récupérables par drainage naturel.* Cours Ecole Nationale Supérieure du Pétrole et des Moteurs (ENSPM), tome V, Editions Technip, Paris, 1968, p. 153 *et sec.*

The flow capacities in the x and y directions are adjusted as follows:

$$(kh)_{n+1} = \frac{(kh)_n \, (h)_{n+1}}{(h)_n}$$

The adjustment of the pressure at a given point is a function of the thickness and of the pressure itself. An iterative process must be used:

$$p_{n+1} = p_n \left(\frac{Z_{n+1}}{Z_n}\right)\left(\frac{h_n}{h_{n+1}}\right)$$

where

$Z_{n+1} = f(p_{n+1})$,

$Z_n = f(p_n)$.

c. *Calculation of limiting water-free production rates*

At each time step the flowrate for each well is calculated on the basis of a fixed pressure gradient between the base of the perforations and the gas-water contact, this gradient being chosen so that water coning will not be initiated.

2. Calculation of swept area

The area swept is calculated by following a fixed number of points (40 in this case) situated on the interface between the dry gas and the reservoir gas.

The velocity of each point on the front is interpolated from the grid of x and y velocities for the centre of each cell.

The frontal displacement calculations are made for short periods of injection of from 1 to 3 years. The fluid displacement velocities are calculated from the pressure distributions obtained for the end of each time step. The injection period is further divided into short time steps such that the movement of any point during a time step does not exceed a given fraction of the blocksize:

$$t = \alpha(\Delta x)/v_{\max} \text{ with } v_{\max} = \sqrt{v_x^2 + v_y^2}$$

The new position of the points is calculated and the swept area obtained.

Breakthrough is detected by comparing the locations of the points with the production well locations.

Using a plotter a visual representation of the positions of the fronts at various time steps may be obtained.

B. Computer time required

For a 25 x 137 grid model, a two year production forecast took 918 seconds of CPU time on a CDC 6 600 computer. The time required to calculate the areas swept was about the same. The grid blocks used were 203 m square.

C. Simulation results

The principal results of these simulations have already been mentioned in Chapter 5. They were, briefly:

Natural depletion study

Two aquifer sizes were studied, $r_e/r_w = 5$ and 9. To provide the required level of production (1.6 million m^3/day) 10 wells were necessary at the start and 22 wells at the end of the development. Final recovery was of the order of 65%.

Recycling study

The model provided the information necessary:

(a) To determine well locations (see Section 54).

(b) To select a rate of replacement of extracted fluids (see Section 51).

(c) To evaluate the relationship between recovery factor and field production rate. Given that the characteristics of the reservoir were relatively unknown, simulations were performed originally for rates of 1.12 and 1.68 million m^3/day, then later for rates of 2.8 and 5.9 million m^3/day. The recovery factor appeared to increase with increasing rate, but this conclusion is only tentative since the calculations were stopped at the end of recycling. Recovery during the blow-down phase after recycling was thus not taken into account.

(d) To study a well location for LPG storage.

(e) To study the problems concerning the concession boundaries.

Some of the results are shown on Fig. A.5.2.

III. MATHEMATICAL MODEL STUDY OF UNITS 2 AND 3

A. Description of the model

Given that units 2 and 3 were a major part of the field (57 wells), a considerable effort was made to obtain an adequate simulation model.

To start with, the model had to be able to cope with three phases: dry gas, wet gas and water. Even though the dry and wet gases were miscible they had to

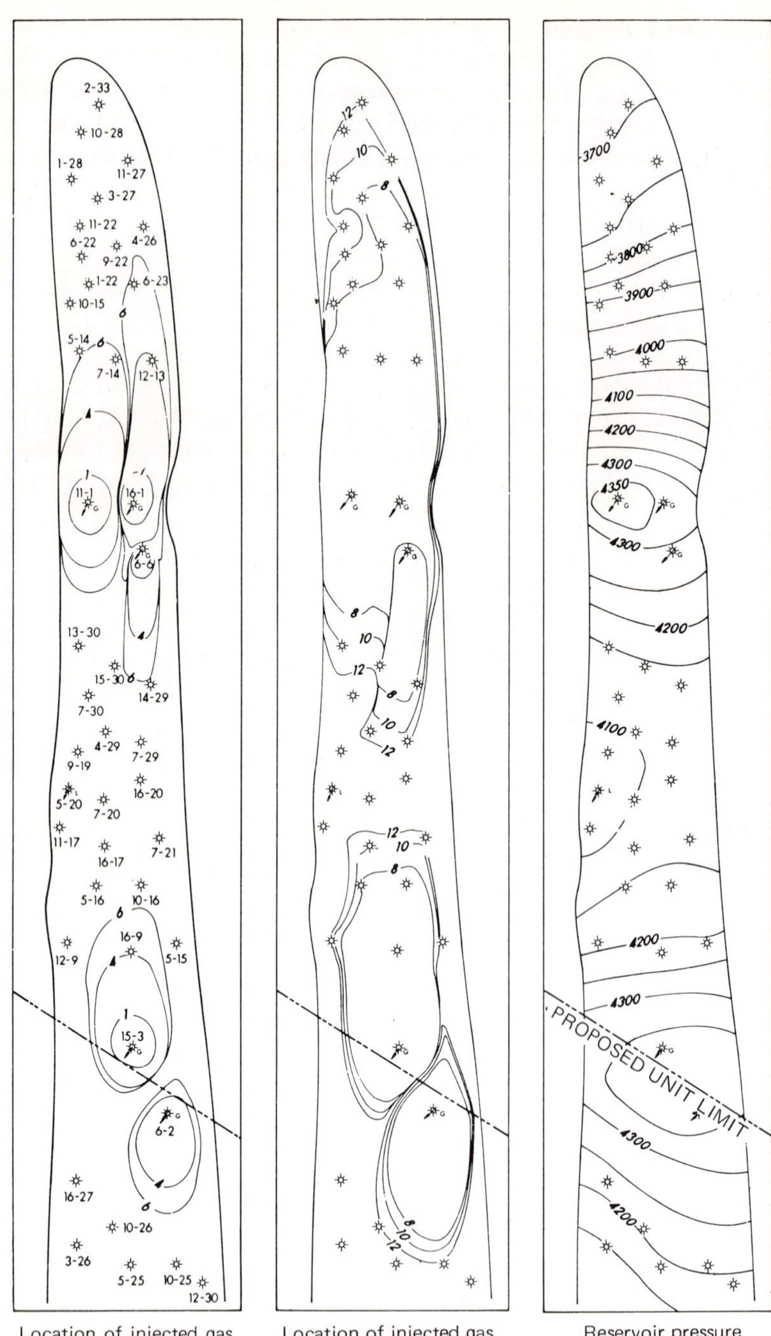

Location of injected gas front after 1, 4 and 6 years.

Location of injected gas front after 8, 10 and 12 years.

Reservoir pressure after 6 years (in psi).

Fig. A.5.2. Results of the gas recycling simulation for unit 1 of the South Kaybob field (Ref. 4).

be treated as separate phases for the purpose of calculation. Retrograde condensation was not taken into account since it was expected that the reservoir pressure would not fall far below the dew-point.

The necessary degree of geometrical complexity was not as easy to define. The presence of the shale bed did not necessarily mean that a three dimensional model would be required, unless there was no communication between the two reservoirs in which case the frontal velocities would be different. The vertical communication across the shale was tested using several monophasic cross-sectional models, the results of which, together with the heterogeneous nature of the formation permeability, led to the conclusion that a three dimensional model was required.

Block size was studied carefully, since finite difference models are generally fairly sensitive to this parameter. The choice made was based on a comparison of the areal sweep efficiency obtained using several areal (2D) models in each of which the block size was different. Four block sizes were examined, the final choice being a 565 m square.

The Kaybob field aquifer is common to two other fields and so a regional study was made using a two phase model with 1 300 blocks. The results from this model were used to define the water influx for the Kaybob field.

Gravity effects and pseudo-relative permeability curves, which enable the treatment of miscible displacement by an immiscible model, were studied in detail.

Finally an independent model was used to study the problem of water coning and to define the limiting pressure gradients to be imposed.

These studies resulted in the choice of a three dimensional, three phase model of 60 x 16 x 2 = 1 920 blocks. The model incorporated the well productivity equation:

$$Q = C(P_c^2 - P_{wf}^2)^n$$

where

P_c is the average pressure of the cell containing the well and

P_{wf} is the bottom-hole flowing pressure of the well based on a pressure gradient of 0.023 bar/m.

B. Results of the model study

The model was used to study various possible injection well configurations (see Section 54) and to predict the production performance, dry gas fraction, number of wells required and areal sweep for each unit. The principal results are shown on Figs. A.5.3 to A.5.8.

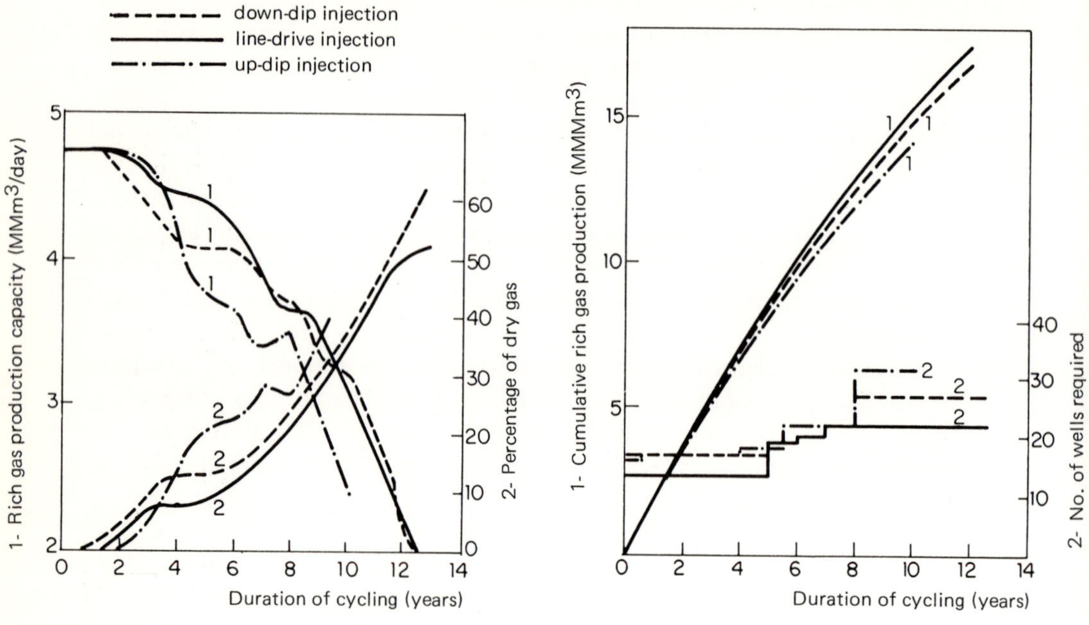

Fig. A.5.3. South Kaybob field.
Performance of unit 2 under gas recycling.

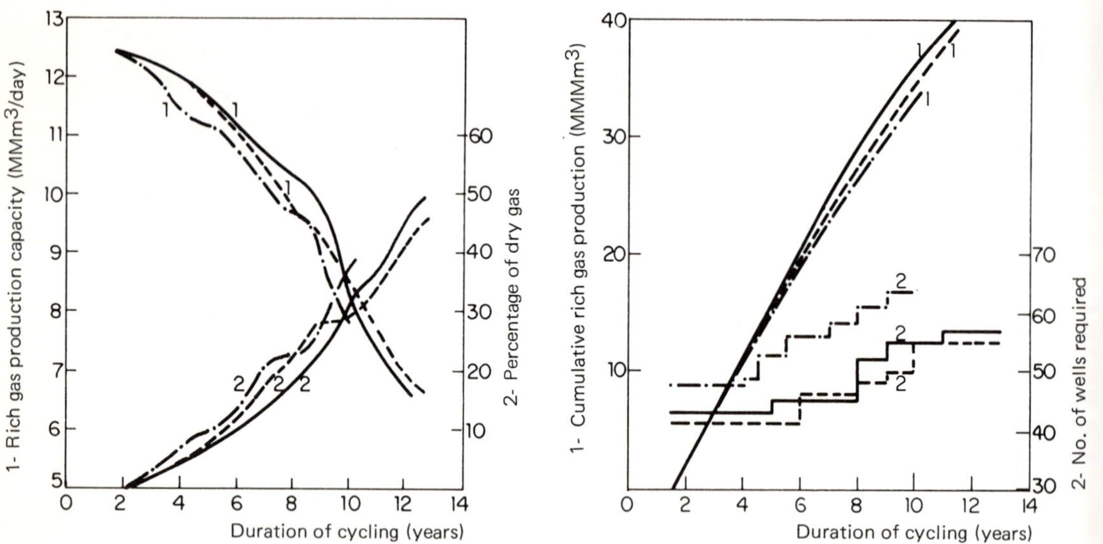

Fig. A.5.4. South Kaybob field.
Performance of unit 3 under gas recycling.

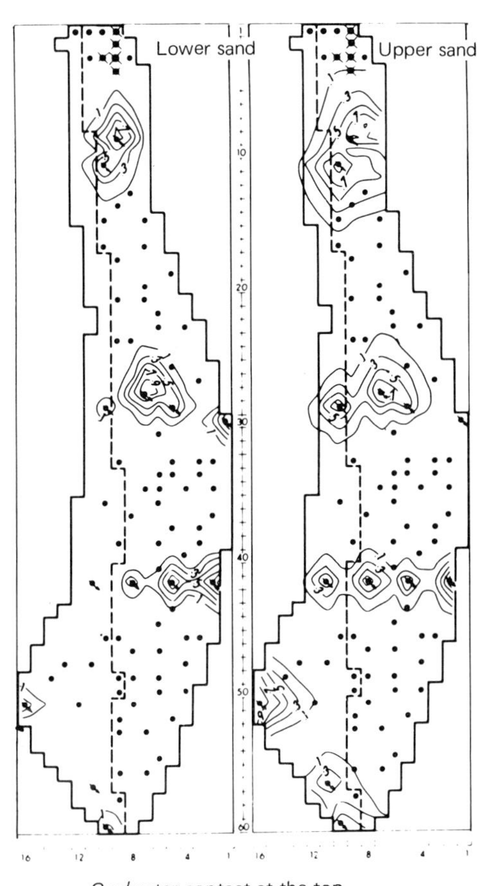

Fig. A.5.5. Line drive injection, dry gas saturation after 3 years (Ref. 5).

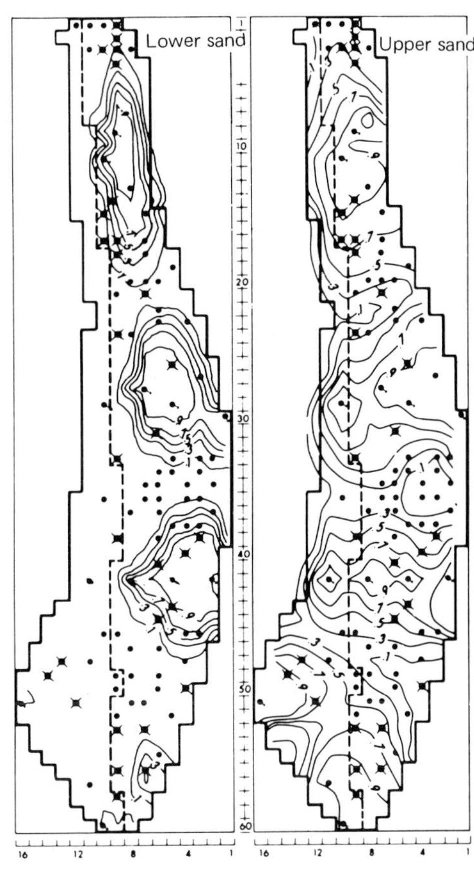

Fig. A.5.6. Line drive injection, dry gas saturation after 10 years (Ref. 5).

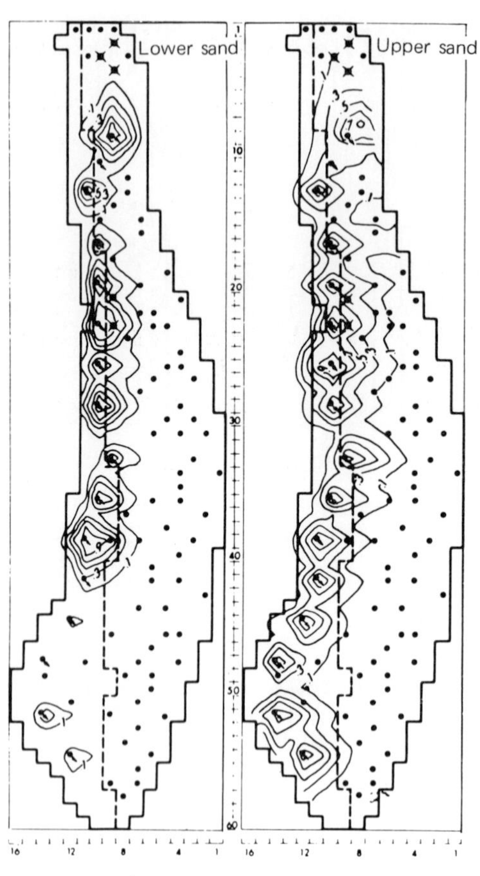

Fig. A.5.7. Downdip injection, dry gas saturation after 3 years (Ref. 5).

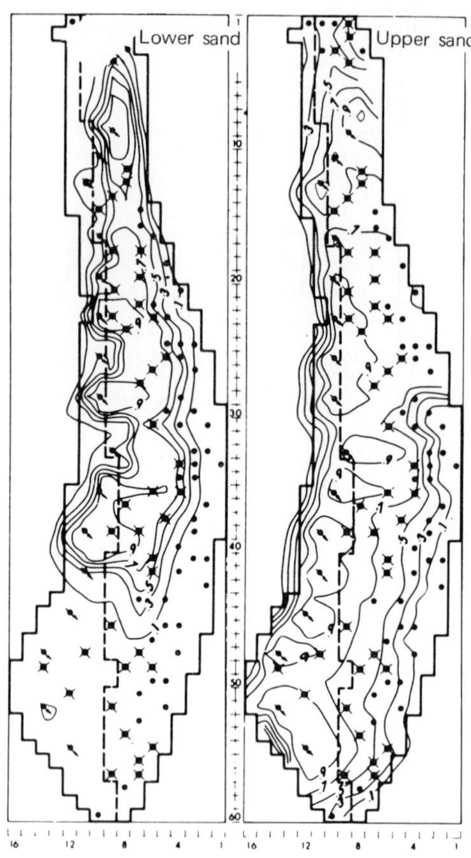

Fig. A.5.8. Downdip injection, dry gas saturation after 10 years (Ref. 5).

REFERENCES

1 CHAUMET, P. and LANCHON, M., "Les réservoirs de gaz à condensat". Etude bibliographique IFP, Ref. 6 936, November 1961.

2 STANDING, M.B., LINDBLAD, E.N. and PARSONS, R.L., "Calculated recoveries by cycling from a retrograde reservoir of variable permeability". *Petrol. Tech.,* vol. 10, May 1947.

3 FRICK, T.C., *Petroleum Production Handbook. Volume II. Reservoir Engineering.* McGraw-Hill Book Company, 1962.

4 FIELD, M.B., GIVENS, J.W. and PAXMAN, D.S., "Kaybob South. Reservoir simulation of a gas cycling project with bottom water drive". *JPT,* April 1970, p. 481-192.

5 FIELD, M.B., WYTRYCHOWSKI, I.M. and PATTERSON, J.K., "A numerical simulation of Kaybob South gas cycling projects". *JPT,* October 1971, p. 1 253-1 262.

6 BRUT, J.J., "La déshydrogénation et le dégazolinage du gaz d'Hassi-R'Mel". *Rev. Inst. Franç. du Pétrole,* January 1962.

7 LEWIS, J.O., "Interpretation of Well Test Data in Gas Condensate Fields". *JPT,* July 1946.

6

thermal recovery methods

by J. BURGER and P. SOURIEAU

In the introduction to this book it was noted that the difference between thermal recovery methods and other recovery methods lies in the fact that the injected fluid supplies thermal energy to the reservoir. There are two categories of thermal methods: those in which the heat is produced at surface (hot fluid injection) and those in which the heat is created in the formation (in-situ combustion). In the first case the injected fluid carries the heat produced while in the second case the injected fluid is one of the reactants involved in an exothermic reaction taking place in the reservoir. There is a very basic difference between the two methods, since the heat supplied tends to flow away from the heated zone and it is evident that the heat lost will be much greater in the first case than in the second.

In the first case the injected fluids, at their maximum temperature, initially come into contact with the swept zone and there is consequently a significant heat loss. In the second case, heat is only released exactly where it is required, i.e. where the oil is to be displaced. Thus the application of hot fluid injection is essentially dependent on its thermal efficiency. This depends on the heat losses both from the injection well-bore to the surrounding formations and from the reservoir to the cap and base rocks. By comparison, heat loss is not generally a limiting factor for in-situ combustion methods.

The thermal efficiency of the various methods may be improved by the recovery of some of the heat trapped in the formation or in the surrounding rocks. For instance, the forward combustion process can be improved by the simultaneous injection of air and water. Other possibilities include the successive application of different methods, for example cold water injection after a partial sweep by steam or by in-situ combustion.

The common factor in all thermal methods is the increase in temperature of part of the reservoir. This involves specific mechanisms which improve both displacement and sweep efficiency, and which increase the rate of production.

We must therefore consider at the outset the influence of temperature on the rock and fluid properties and thus on the dynamic behaviour of the fluids; the chemical processes involved in thermal methods will also be briefly considered. Hot fluid injection and in-situ combustion will then be discussed in turn. Further details of the methods presented are to be found (a book to be published by *Éditions Technip*) (Ref. 1).

61. DATA REQUIRED FOR THE STUDY OF THERMAL RECOVERY METHODS

61.1. The effects of temperature on hydrodynamic fluid properties

According to Darcy's law, the flow of a multiphase fluid in a porous medium is directly proportional to the relative permeability to the fluid and inversely proportional to its viscosity. These two parameters are very temperature dependent.

A. Viscosity

1. Liquid viscosity

The viscosity of a liquid falls markedly with increasing temperature. The relationship is of exponential form, and the higher the viscosity of the fluid the greater the reduction in viscosity for a given temperature increase.

An example of the variation of viscosity with temperature for a particular crude oil is given in Fig. 61.11., and an ASTM standard viscosity-temperature plot for several crudes is shown in Fig. 61.12. It can be seen, for example, that a crude oil with a viscosity of about 50,000 cSt at 40°C has a viscosity of less than 20,000 cSt at 50° C. By comparison, an oil of 9 cSt viscosity at 40° C still has a viscosity of more than 6 cSt at 50° C.

The dynamic viscosity of liquid water at 20° C is 1 cP, and between 40 and 50° C its viscosity only changes from 0.65 to 0.55 cP.

It can therefore be appreciated that even a small increase in temperature in a formation containing both water and viscous oil leads to a significant reduction in viscosity contrast, favouring the flow of oil rather than water. This explains the interest in thermal recovery methods for the exploitation of viscous oil reservoirs. The decrease in the ratio μ_o/μ_w as temperature increases is shown for two crude oils in Fig. 61.13. Certain light oils exhibit an inverse relationship (dashed line on Fig. 61.13) but this is fairly rare (Ref. 2).

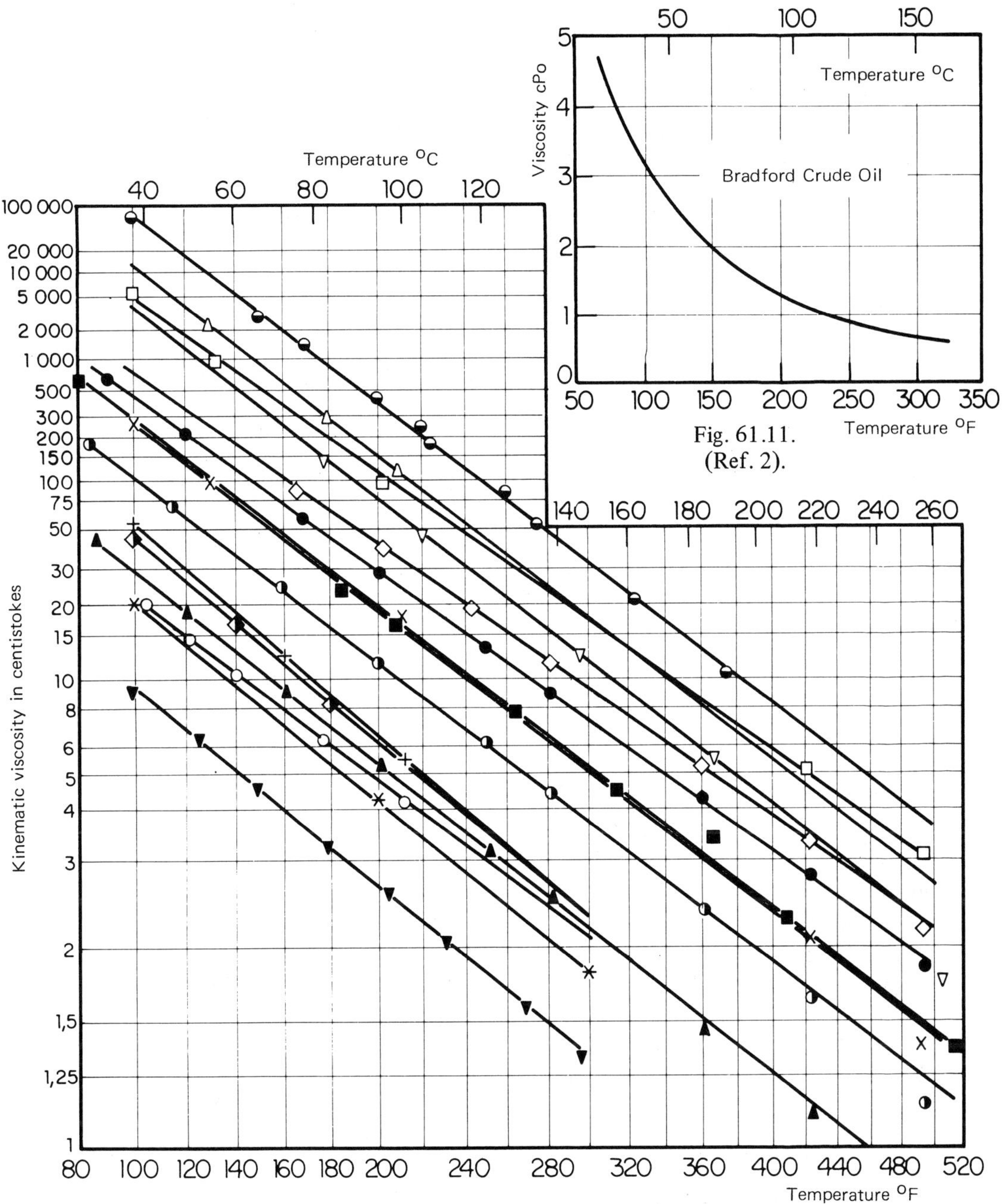

Fig. 61.11.
(Ref. 2).

- ◒ Californian crude
- △ Colombian crude
- □ Mid-Continent residue
- ▽ Californian crude
- ◇ Mid-Continent cylinder oil
- ● Mid-Continent motor oil
- ■ Gulf Coast crude
- X North Louisiana heavy oil
- ◑ Mid-Continent red oil
- + South-Texas crude
- ▲ Gulf Coast crude
- ▲ Mid-Continent light paraffin oil
- ○ Pennsylvania vacuum distillate
- ✳ Wyoming crude
- ▼ Mid-Continent press oil

Fig. 61.12.
(Ref. 3).

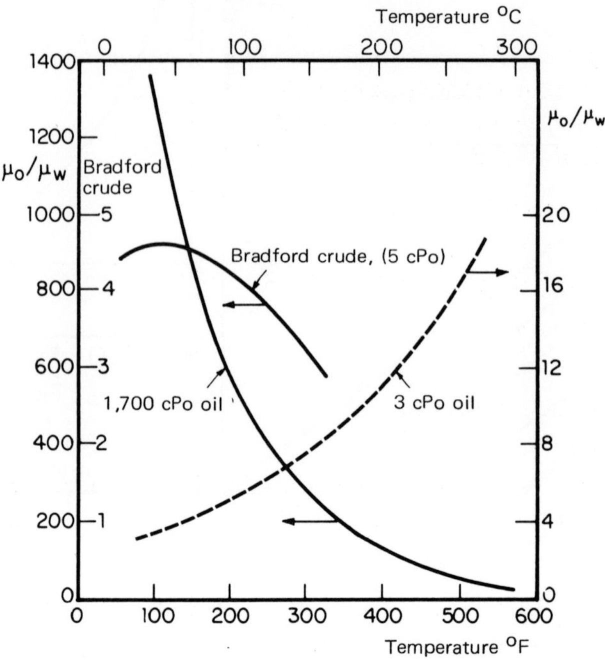

Fig. 61.13.
(Ref. 2).

It should be noted that the viscosity of liquid hydrocarbons may be reduced by the absorption of certain gases, the effect increasing with the quantity of gas dissolved. This viscosity reduction results from the swelling of the oil by the dissolved gas. Thus the absorption of carbon dioxide formed during in-situ combustion may appreciably increase the oil mobility, assuming that the reservoir pressure is sufficiently high that a large quantity of CO_2 is dissolved.

The following formulae are often used in the calculation of the effect of temperature on the viscosity of liquids:

- Andrade's exponential formula:

$$\mu = A \exp(B/T) \qquad \text{(Eq. 61.11)}$$

where

μ is the dynamic viscosity in centipoise,
T is the absolute temperature in K,
A and B are constants.

- Braden's formula for oils (Ref. 3):

$$\log(\nu_2 + C) = \left(\frac{T_1}{T_2}\right)^D \log(\nu_1 + C) \qquad \text{(Eq. 61.12)}$$

where

the subscripts 1 and 2 refer to the absolute temperatures T_1 and T_2,
ν is the kinematic viscosity in centistokes,
C is a constant (equal to 0.6 if $\nu > 1.5$ cSt),
D is a constant of the order of 3.5 to 4.

• Bingham's formula for water:

$$\frac{1}{\mu_B} = 0.021482[(\theta - 8.435) + \sqrt{8\,078.4 + (\theta - 8.435)^2}] - 1.2 \qquad \text{(Eq. 61.13)}$$

where

μ_B is the viscosity in centipoise,
θ is the temperature in ° C.

This last formula requires correction near the critical point (see Fig. 61.14), and it is suggested that the following formula be used if the temperature is greater than 160° C:

$$\mu_e = \mu_B + 7 \times 10^{-9}(\theta - 160)^2 (351 - \theta) \qquad \text{(Eq. 61.14)}$$

where

μ_e is the viscosity of water at $\theta > 160°$ C, centipoise,
μ_B is the viscosity of water from Bingham's formula, centipoise,
θ is the temperature in ° C.

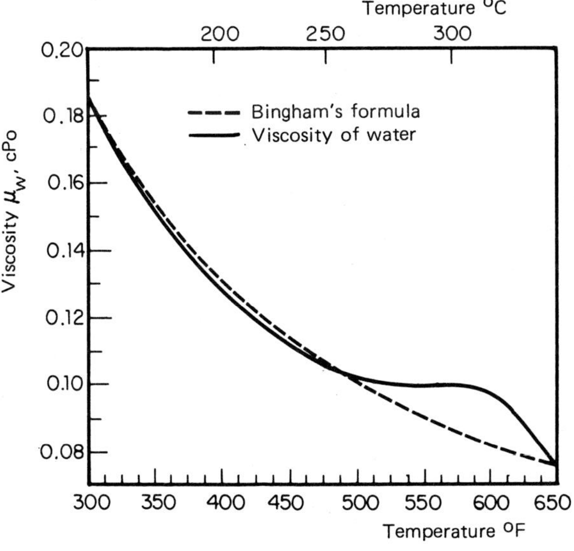

Fig. 61.14.
(Ref. 2).

2. *Gas viscosity*

According to the kinetic theory of ideal gases, the dynamic viscosity of a gas should be independent of pressure and proportional to the square root of the absolute temperature. However, real gases depart from the theory to some degree: gas viscosity tends to increase with pressure and normally increases more rapidly than the square root of the absolute temperature, T. The effect of T on μ can be approximated by the following equation:

$$\mu = AT^n \qquad \text{(Eq. 61.15)}$$

The exponent n lies between 0.7 and 1 for many gases. For example, from Ref. 2 we have:

Steam between 0° and 400° C: $\mu = 1.7 \times 10^{-5}\, T^{1.116}$ cP
Methane between 0° and 500° C: $\mu = 1.36 \times 10^{-4}\, T^{0.77}$ cP

For certain gases linear relationships can be found to express the viscosity as a function of temperature in ° C:

Steam	$\mu = 88 + 0.38\,\theta$ μP	$0 < \theta < 700°$ C
Air	$\mu = 175 + 0.38\,\theta$ μP	$0 < \theta < 500°$ C
Nitrogen	$\mu = 170 + 0.38\,\theta$ μP	$0 < \theta < 400°$ C
Carbon Dioxide	$\mu = 145 + 0.38\,\theta$ μP	$0 < \theta < 500°$ C

These equations are valid for pressures below 100 atmospheres.

B. *Relative permeability*

Various experimenters have shown that relative permeability varies with temperature for diphasic oil-water flow (Refs. 4, 5 and 6). Examples of relative permeability curves at two different temperatures are shown in Fig. 61.15.

It can be seen that when the temperature is increased the irreducible water saturation S_{wi} increases, whereas the residual oil saturation S_{or} decreases (Fig. 61.16). This clearly illustrates that an improved displacement efficiency can be obtained by thermal recovery methods.

The effect of temperature on S_{wi} and S_{or} is the result of both the reduction in the viscosity ratio μ_o/μ_w as temperature increases and changes in the physical and chemical equilibrium within the porous medium.

Indeed, an approximate correlation has shown that S_{or} increases with increasing μ_o/μ_w (Ref. 7). In addition, it has been observed that the relative permeability curves are hardly influenced by temperature in the case of a water-tetradecane system, for which the ratio μ_o/μ_w is fairly insensitive to temperature (Fig. 61.17), (Ref. 6).

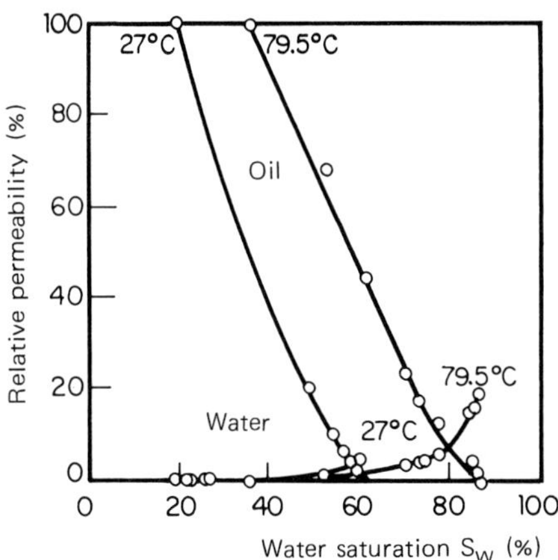

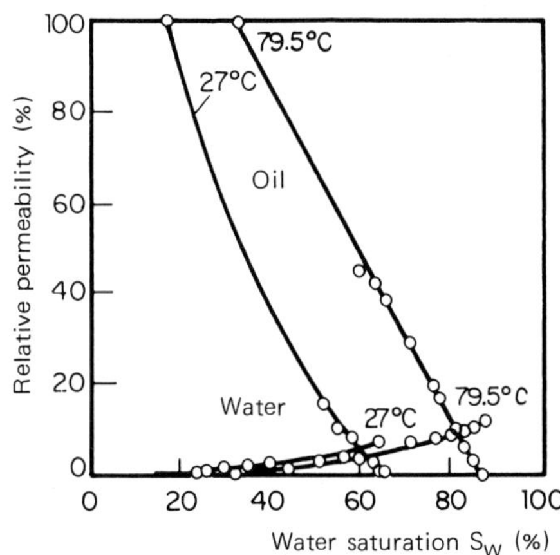

Fig. 61.15.

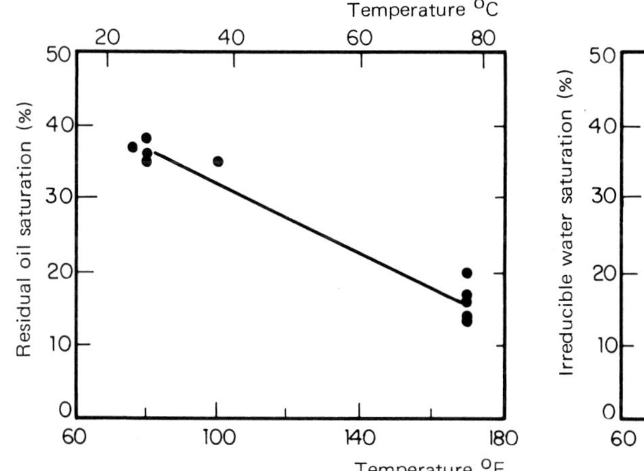

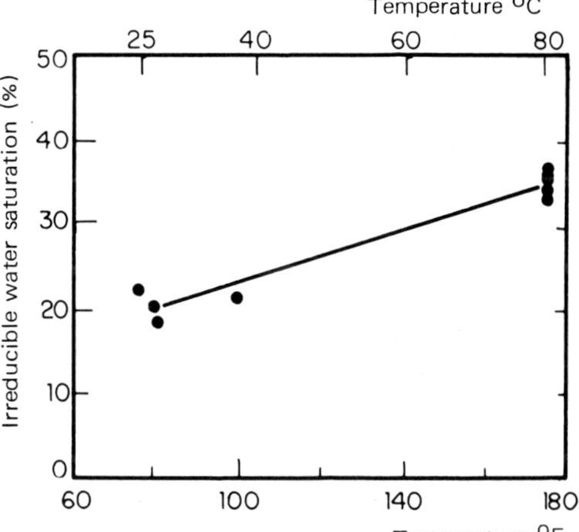

Fig. 61.16.

Figs. 61.15. and 61.16. Boise sandstone cores.
White oil, $\mu_0 \approx 60$ *cP.* at 40°C.

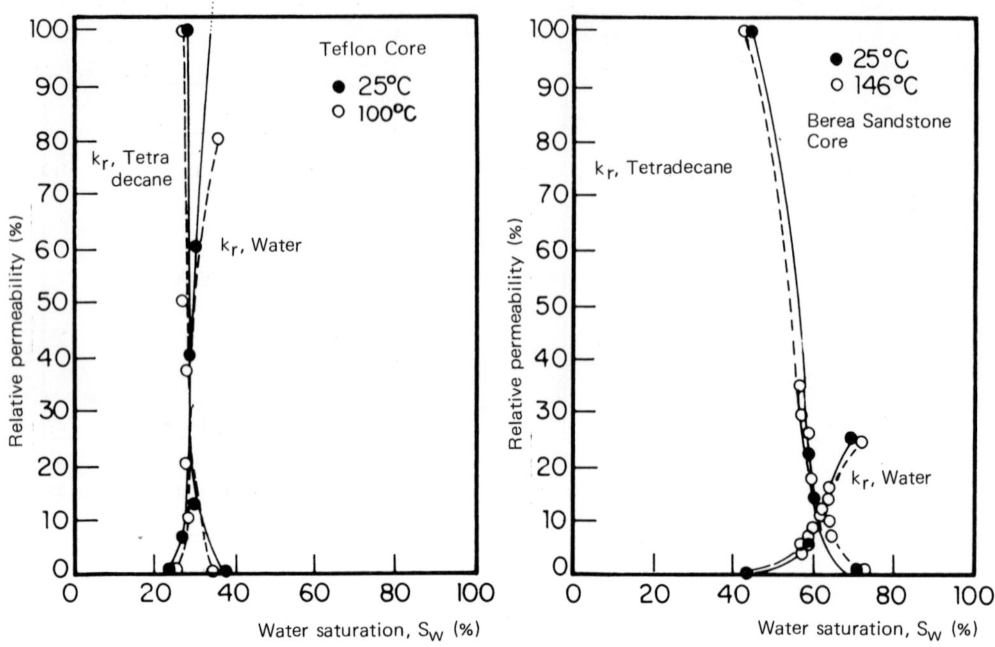

Fig. 61.17.
(Ref. 6).

However, viscosity is not the only parameter to be considered since rock, far from being an inert solid, has an influence due to its adsorptive properties, and oils are complex hydrocarbon mixtures which may exhibit polar behaviour. Clearly fluid interfacial tension and rock wettability vary when the temperature increases; it appears that they both change in the sense of a reduction in oil saturation: the interfacial tension decreases by 10 to 20% for a temperature increase of 60° C over ambient temperature, and the contact angle of an oil-water system on glass changes in the direction of better wettability to water. However, caution is still necessary since, for example, during the displacement of oil by the hot water downstream from an in-situ combustion front, the presence of oxidized products from the reaction between the injected oxygen and the hydrocarbons in place may render the rock more oil-wet while decreasing the interfacial tension.

The variation of the ratio k_{ro}/k_{rw} with temperature is less predictable than the variation of the residual oil saturation. Considerable change in this ratio has been observed, but the direction of the change cannot be predicted (Ref. 4).

In the more complex case where a gaseous phase is present, the phenomenon of vaporization-condensation may be added to the hydrodynamic displacement represented by the generalized form of Darcy's law. There may be large heat and mass transfers which principally affect the lightest fractions of the oil. This is the reason why the residual oil saturation is lower for displacement by saturated steam than for displacement by liquid water at the same temperature (but at higher pressure such that no vapour phase is formed).

The effect of the vaporization-condensation phenomenon, which has been observed both in the laboratory and in the field, is larger as the content in volatile components of the oil increases.

61.2. The effects of temperature on the thermal and thermodynamic properties of fluids and solids

A. Thermal expansion

The value of the coefficient of cubic expansion is of the order of 10^{-3}/°C for oil and 3 x 10^{-4}/°C for water. For rock, the coefficient is of the order of 10^{-5}/°C. An increase of temperature thus tends to encourage the expulsion of oil from the pore space.

B. Thermal capacity

Generally speaking, the thermal capacity per unit mass (or specific heat) of solids, liquids and gases increases with temperature and is often affected by pressure. For perfect gases, the specific heat is independent of pressure but may increase with temperature due to progressive molecular excitation.

The influence of the absolute temperature, T, on specific heat is often expressed in the form:

$$c_p = A + BT + CT^2 \tag{Eq.61.21}$$

where

c_p is the specific heat at constant pressure, and

A , B and C are constants.

For petroleum fractions and oils in the liquid state, the following approximate equation can be used (Ref. 2):

$$c_p = (0.403 + 0.00081\theta)/\sqrt{d} \text{ cal/(g° C)} \tag{Eq.61.22}$$

where

θ is the temperature in °C and,

d is the specific gravity at 15° C.

Note that the thermal capacity of crude oils is of the order of 0.45 cal/(g° C) at ambient temperature, being less than half of the thermal capacity of water.

The specific heat of rocks varies slightly according to their nature (Ref. 8): at ambient temperature, the value of c_p is between 0.18 and 0.21 cal/(g° C) for sand and natural rocks in the dry state. For these solids, the value of c_p at a temperature θ (° C) may be expressed by the following average relationship:

$$c_p \approx 0.2 + 1.8 \times 10^{-4}\ \theta \text{ cal/(g° C) } 0 \leqslant \theta \leqslant 500°\text{C}$$

Thermal capacities are additive, thus the equivalent specific heat per unit volume $(\rho c)^*$ of a medium of porosity ϕ, of which the supporting mineral s contains i fluids with pore volume saturations S_{f_i}, is given by:

$$(\rho c)^* = (1 - \phi)\,(\rho c)_s + \phi \sum_i S_{f_i}\,(\rho c)_i$$

C. *Thermal conductivity*

The thermal conductivity of gases increases with temperature whereas that of most liquids and solids decreases slightly with temperature. Water is however an exception: its conductivity passes through a maximum at around 130° C. The conductivity of hydrocarbons at ambient temperature is of the order of 0.12 kcal/(mhr° C), while that of water at 20° C is 0.51 kcal/(mhr° C). The thermal conductivity of rocks is a few kilocalories per mhr° C.

For a porous medium saturated with given fluids, we can define an equivalent thermal conductivity λ^* by simulating the composite medium with an imaginary continuous medium, with the condition that the temperature of the fluids is close to that of the rock. However, given that the conductivities are not additive, the equivalent conductivity depends not only on the porosity and the saturations, but also on the structure of the porous medium, which effectively determines the available area for heat transfer from point to point. Various models of porous media have been studied in order to attempt to estimate the equivalent conductivity of a medium as a function of the properties of its constituents (Ref. 1.). Thus for a medium of porosity ϕ consisting of a rock of conductivity λ_s and fluids of conductivity λ_{f_i} present in the pores with saturations S_{f_i}, we may, for example, use the following approximate equation:

$$\lambda^* = \lambda_s^{(1-\phi)} \left[\sum_i \lambda_{f_i} S_{f_i}\right]^{\phi} \qquad \text{(Eq.61.23)}$$

It should be noted that the errors in estimates made using an equation of this form depend on the structure of the porous medium and the distribution of the fluids.

D. *Latent heat of vaporization*

The temperature at which a substance vaporizes is a function of its pressure p. For instance, the vaporization temperature θ_{vap} of water is given by the following equation for $p > 0.7$ atm:

$$\theta_{\text{vap}} = 130\,p^{0.21} - 30°\text{ C } (p \text{ in atm})\ ;\ p > 0.7 \text{ atm}$$

The latent heat of vaporization Q_v of a pure substance varies with the temperature at which the phase change takes place: Q_v decreases as T increases and becomes zero at the critical point (Fig. 61.21). Water is the substance with the highest latent heat of vaporization, being 583.2 cal/g at 25° C and 539 cal/g at 100° C. Most hydrocarbons have a latent heat of vaporization of around 80-90 cal/g at 25° C.

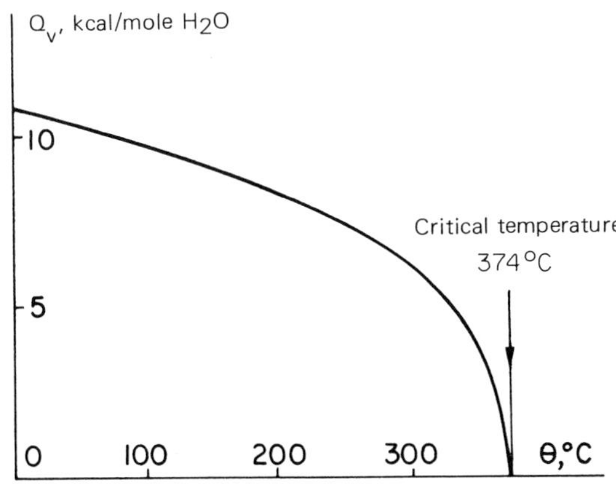

Fig. 61.21.

61.3. Chemical reactions occurring during thermal recovery

Certain thermal recovery methods benefit from the reactivity of the hydrocarbons constituting the crude oil, particularly in respect of oxidation and combustion. In this section the basic thermodynamic and kinetic reactions will be discussed.

A. The thermodynamics of chemical reactions

Two state variables are of particular significance during a chemical reaction: the enthalpy and the Gibbs function (or free enthalpy).

1. Enthalpy of reaction

Consider a chemical transformation corresponding to the following stoechiometric balance:

$$\alpha A + \beta B \rightarrow \gamma C + \delta D$$

The accompanying change in enthalpy, or enthalpy of reaction ΔH°, determined at a standard reference temperature, is a combination of the standard enthalpies of formation ΔH_f° of the reactants and products:

$$\Delta H^\circ = \gamma\ \Delta H_{fC}^\circ + \delta\ \Delta H_{fD}^\circ - \alpha\ \Delta H_{fA}^\circ - \beta\ \Delta H_{fB}^\circ$$

The enthalpy of reaction is negative for exothermic reactions and positive for endothermic reactions. The oxidation and combustion of hydrocarbons are strongly exothermic reactions: complete combustion to CO_2 and water liberates around 11,000 kcal/kg of fuel.

TABLE 61.31.

Reactant	Oxygen	Products	$[H_2O\ \text{liq.}]$ kcal/mole O_2
$-\overset{(H)}{\underset{(H)}{C}}-H$	$+\frac{3}{2}O_2$	$\to CO_2 + H_2O$ $\to -C\begin{smallmatrix}=O\\ -OH\end{smallmatrix} + H_2O$	≈ 105
	$+O_2$	$\to CO + H_2O$ $\to -C\begin{smallmatrix}-H\\ =O\end{smallmatrix} + H_2O$	≈ 90
	$+O_2$	$\to -\overset{\vert}{\underset{\vert}{C}} = O + H_2O$	90-100
	$+O_2$	$\to -\overset{\vert}{\underset{\vert}{C}} - O - O - H$	≈ 30
	$+\frac{1}{2}O_2$	$\to -\overset{\vert}{\underset{\vert}{C}} - OH$	70-90

In practice, hydrocarbons may be transformed into a wide variety of oxidization or combustion products when they react with oxygen (e.g. aldehydes, ketones, alcohols, carbon monoxide, carboxylic acids, CO_2 etc.) (Ref. 9). All these reactions are very exothermic: the heat liberated is between 90 and 105 kcal/mole of oxygen consumed (see Table 61.31). Only the transient formation of peroxides is relatively weakly exothermic (of the order of 30 kcal/mole oxygen).

2. *Gibbs function*

Certain reactions are reversible, according to the following scheme:

$$\alpha A + \beta B \rightleftarrows \gamma C + \delta D$$

If the compounds A, B, C and D are brought to a given temperature T, the reaction occurs either in one direction or the other until thermodynamic equilibrium is attained. The equilibrium concentrations $(J)_{eq}$ of the J components are related by the following equation:

$$k_{eq} = \frac{(C)_{eq}^{\gamma}\,(D)_{eq}^{\delta}}{(A)_{eq}^{\alpha}\,(B)_{eq}^{\beta}} \qquad \text{(Eq. 61.31)}$$

The constant k_{eq} is independent of the concentrations $(J)_{eq}$ of the various components. If the value of k_{eq} is calculated by expressing the concentrations in the form of partial pressures in the mixture at equilibrium we have:

$$\ln k_{eq} = -\frac{\Delta G^{\circ}}{RT} \qquad \text{(Eq. 61.32)}$$

where

ΔG° is the standard free enthalpy of the reaction (Gibbs function): $\Delta G^{\circ} = \Delta H^{\circ} - T\Delta S^{\circ}$,

R is the universal gas constant,

T is the absolute temperature.

When the free enthalpy of formation ΔG_f° of a compound is positive it should in principle tend to decompose, since the equilibrium constant favours the pure constituents of which it is made. However, a possible thermodynamic transformation will only take place in practice if the reaction rate is sufficiently high (most hydrocarbons, despite being fairly stable, have positive free enthalpies of formation).

Let us consider the following hypothetical reaction between two hydrocarbons 1 and 2:

$$\frac{1}{n_1} C_{n_1} H_{m_1} \rightleftarrows \frac{1}{n_2} C_{n_2} H_{m_2} + \left(\frac{m_1}{2\,n_1} - \frac{m_2}{2\,n_2}\right) H_2$$

The free enthalpy of this reaction is:

$$\Delta G^{\circ} = \frac{\Delta G_{f_2}^{\circ}}{n_2} - \frac{\Delta G_{f_1}^{\circ}}{n_1}$$

If $\Delta G_{f_1}^{\circ}/n_1 > \Delta G_{f_2}^{\circ}/n_2$ the formation of hydrocarbon 2 takes place in preference to that of hydrocarbon 1. Values of $\Delta G_f^{\circ}/n$ are shown in Fig. 61.31 for various hydrocarbons. It appears that paraffins are the most stable compounds at low temperatures, but that at high temperatures (from around 300° C) they tend to decompose into less hydrogenated hydrocarbons: aromatics, olefins and acetylenic hydrocarbons. The chemical alteration of hydrocarbons under the effect of temperature (pyrolysis) often leads to the formation of carbonaceous deposits containing little hydrogen.

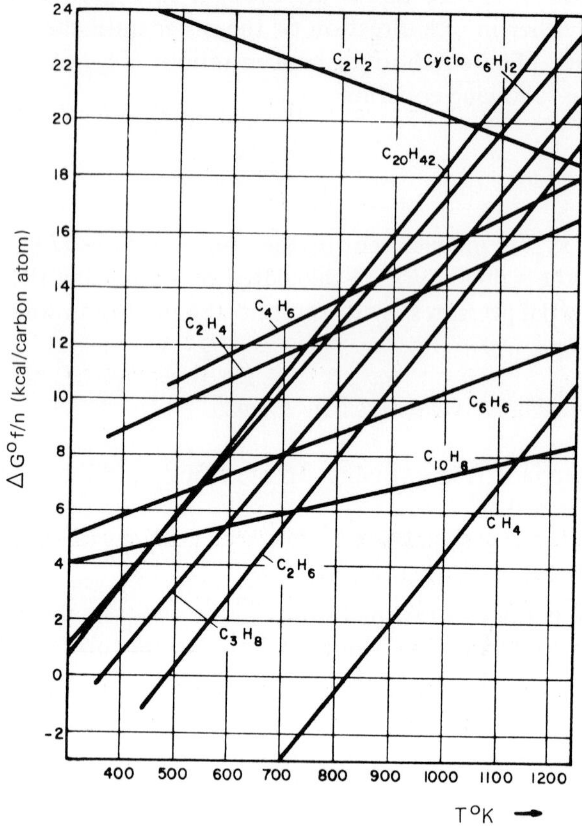

Fig. 61.31.

B. Chemical kinetics

1. Basic principles

If in a volume v_r, the number of moles n_J of a component J change with time due to a chemical reaction, the rate of reaction U_J' is defined as follows:

$$U_J' = \frac{1}{v_r} \left| \frac{dn_J}{dt} \right| \qquad \text{(Eq. 61.32)}$$

The reaction rate may be determined by considering any one of the reactants or products involved in the reaction. For any reaction of the type:

$$\alpha A + \beta B \rightarrow \gamma C + \delta D$$

we have:

$$\frac{U'_A}{\alpha} = \frac{U'_B}{\beta} = \frac{U'_C}{\gamma} = \frac{U'_D}{\delta} = U'$$

The rate of reaction in a process involving no phase change may be expressed as follows:

$$U' = K \prod_J \left[\left(\frac{n_J}{v_r}\right)^{r_J}\right] \qquad \text{(Eq. 61.33)}$$

The exponent r_J is known as the "order" of the reaction in respect of the component J. Values of the exponent can be determined experimentally and provide information about the reaction mechanisms. For simple reactions the exponent is unity. K is the rate constant for the reaction and increases with temperature according to Arrhenius' law:

$$K = K_0 \exp\left(-E/RT\right) \qquad \text{(Eq. 61.34)}$$

where

E is the activation energy, normally expressed in units of kilocalorie/mole or joule/mole,
R is the universal gas constant,
T is the absolute temperature.

In the special case of a surface reaction, the surface area s_r over which the reaction takes place must be considered, and Eq. 61.33 becomes:

$$U' = K \frac{s_r}{v_r} \prod_J \left[\left(\frac{n_J}{v_r}\right)^{r_J}\right] \qquad \text{(Eq. 61.35)}$$

2. Reaction kinetics in porous media

Some chemical reactions can take place in the pores of a porous medium: given the structure of a typical medium these reactions often involve the pore surfaces or the surface of an interface between phases (Ref. 9).

Consider the case of a medium whose pores (volume v_n) contain a saturation S_o of oil of density ρ_o and possibly a saturation in an oxygen-bearing gas. It has already been noted that hydrocarbons are pyrolyzed by the effect of heat and that the reaction of oxygen with hydrocarbons releases a great amount of heat.

In particular the following reactions may occur:

(a) Oxidation of oil droplets by the dissolved oxygen. If the oxygen is uniformly dissolved throughout the droplets (this assumes that the droplets are small and that the temperature is low) the reaction rate, expressed as the mass of oxygen consumed per unit volume v of the porous medium ($v = v_p/\phi$) may be written as follows:

$$U_1' = \frac{1}{v}\left|\frac{dm_{O_2}}{dt}\right| = K_1\,\phi\rho_o S_o P_{O_2}^n \qquad \text{(Eq. 61.36)}$$

where

P_{O_2} is the oxygen partial pressure in the gas present in the pores.

(b) Combustion of a solid carbonaceous deposit: this reaction involves the oxygen reacting on the surface s_d of the deposit. The change in the mass of the deposit may be expressed as follows:

$$\frac{1}{v_p}\left|\frac{dm_d}{dt}\right| = K_2\left(\frac{s_d}{v_p}\right)\left(\frac{m_d}{v_p}\right)P_{O_2}^{n^*}$$

The value of s_d/v_p is a complex function of the surface area per unit pore volume s_p/v_p of the porous medium; it depends on the distribution of the deposit within the pores.

Per unit volume of the porous medium we have:

$$U_2' = \frac{1}{v}\left|\frac{dm_d}{dt}\right| = K_2\,\phi\left(\frac{s_d}{v_p}\right)\left(\frac{m_d}{v_p}\right)P_{O_2}^{n^*} \qquad \text{(Eq. 61.37)}$$

(c) Superficial pyrolysis of the oil droplets (mass m_o, surface area s_o, pore volume v_p) to form a carbonaceous deposit:

$$+\frac{1}{v_p}\frac{dm_d}{dt} = K_3\left(\frac{s_o}{v_p}\right)\left(\frac{m_o}{v_p}\right) = K_3\left(\frac{s_o}{v_p}\right)\rho_o S_o$$

The value of s_o/v_p depends on the distribution of the oil droplets in the pores. Finally we have:

$$U_3' = \frac{1}{v}\frac{dm_d}{dt} = K_3\,\phi\rho_o S_o\left(\frac{s_o}{v_p}\right) \qquad \text{(Eq. 61.38)}$$

Obviously, reaction (b) can only take place after reaction (c) has resulted in the formation of carbonaceous deposit.

62. HOT FLUID DISPLACEMENT

As we have seen in Section 61, water has the advantage of having a much higher heat transport capacity than any other fluid, whether in the liquid or vapour phase. For this reason, water is the only hot fluid ever used.

The quality X of saturated steam is generally defined as the mass of dry steam contained in unit mass of the wet vapour. The enthalpy of the mixture is thus given by:

$$H = H_\ell + XL$$

where

H_ℓ is the enthalpy of liquid water on the saturation curve and,

L is the latent heat of vaporization at constant pressure.

At the vaporization temperature (as long as it is not too close to the critical temperature), dry steam transports far more heat than liquid water. Consider for example (see Fig. 62.1) reservoirs in which the temperature varies according to the normal geothermal gradient of 3°/100 m (based on a surface temperature of 20° C) and in which the injection pressure is equal to the hydrostatic pressure. Recoverable heat is defined as the difference in enthalpy between the hot fluid injected and liquid water at bottom-hole conditions. The ratio between the recoverable heat at $X = 1$ and that at $X = 0$ is 3.4 at 20 bar but still 1.8 at 150 bar.

It should be noted that, there is generally no advantage in using superheated steam. For example, using the above hypotheses, in the neighbourhood of the saturation curve the increase in recoverable energy per °C of superheating is only 0.1 % at 20 bar and 0.2 % at 100 bar. This minor gain does not compensate for the increased mechanical problems involved in operating at the higher temperatures.

During the injection of saturated steam part of the heat is lost into the rocks adjacent to the well-bore, and consequently the steam quality falls. However, as long as the quality is not too low, the recoverable heat remains high. For example, the ratio of recoverable heat for $X = 0.6$ and $X = 1$ increases from 0.72 at 20 bar to 0.83 at 150 bar.

The above discussion seems to imply that, from the point of view of thermal efficiency, the injection of saturated steam is to be preferred to the injection of hot water. We shall see that this is not always the case in practice.

62.1. Basic principles (Refs. 1, 11)

A. Hot water displacement

Consider a virgin homogeneous reservoir undergoing a unidimensional displacement in which heat loss to the surrounding formations is neglected. The injected hot water cools on contact with the rock and fluids in place and, under steady-state conditions, two principal zones can be distinguished on the temperature and saturation profiles. The zones will be described from downstream to upstream (Fig. 62.11).

Recoverable enthalpy
(kcal/kg)
θ (°C)
X=1
0.8
0.6
0.4
0.2
0
Critical point
Hydrostatic pressure
Normal geothermal gradient (3°C/100 m)
Surface temperature 20°C
Pressure (kg/cm2)
600
400
200
0
400
300
200
100
20
40
60
80
100
120
140
160
180
200
220
240

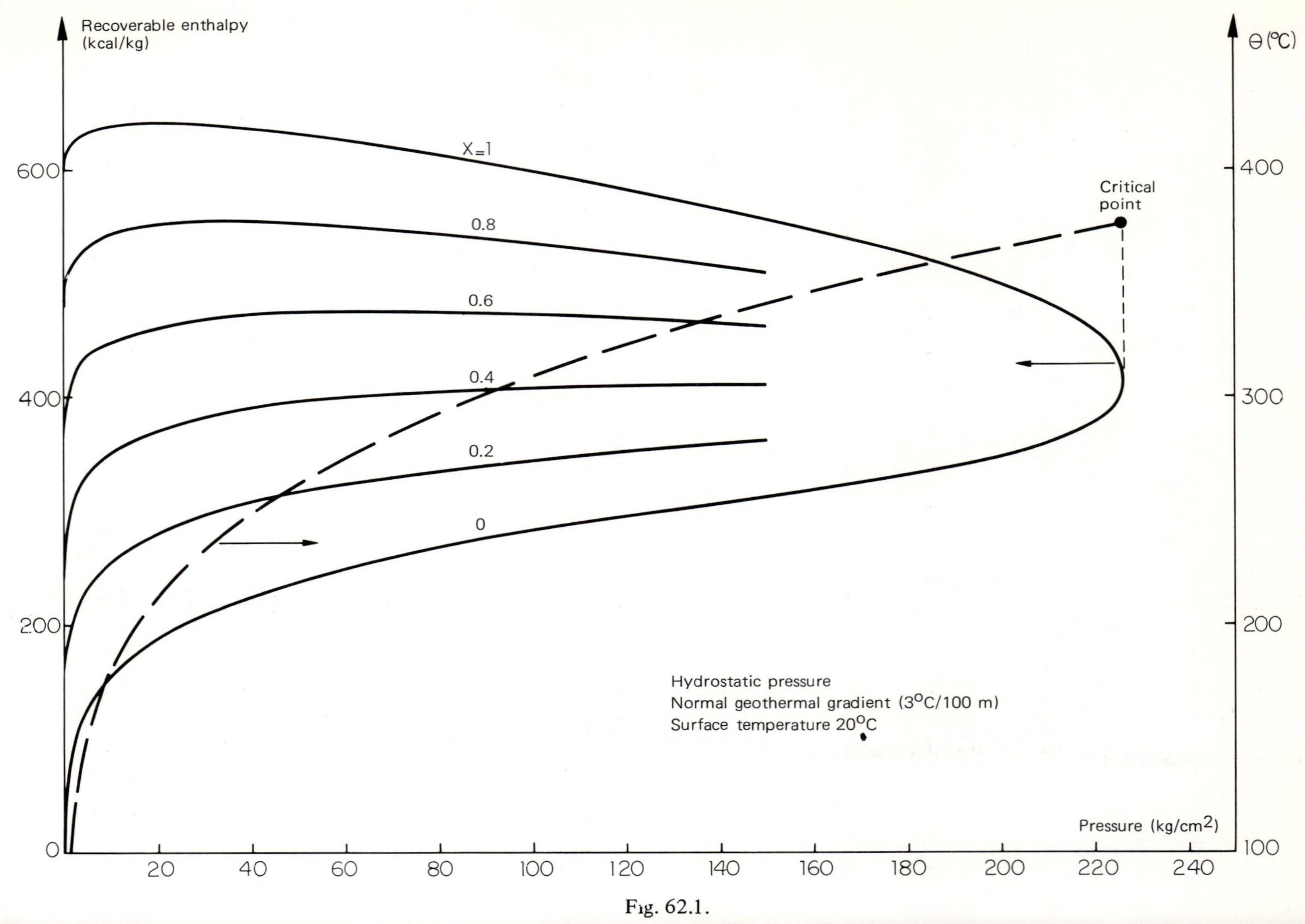

Fig. 62.1.

Zone II. The oil in place is displaced by water at the same temperature. Consequently, displacement by hot water suffers from the same instability problems as that by cold water; the residual oil saturation upstream of zone II is the same as it would be for cold water injection.

Zone I. As we move further upstream the temperature steadily increases while the residual oil saturation decreases as shown in Fig. 61.16. In addition, the swelling of the fluids and matrix means that, for any given saturation, the mass of oil trapped is reduced as the temperature increases. If the oil contains any highly volatile hydrocarbons, certain fractions may be displaced by vaporization-condensation.

In practice, the loss of heat from the hot zone to the surrounding formations results in a more pronounced temperature loss in the direction of flow but does not affect the rate of advance of this zone (Fig. 62.12).

B. Displacement by saturated steam (Ref. 2)

With the same hypotheses as for hot water displacement, three principal zones can be identified (Fig. 62.13), but this time they will be examined from upstream to downstream.

Zone I. Upstream of the condensation zone the temperature is high and falls only slightly according to the saturation temperature of steam at the prevailing pressure, which naturally declines in the direction of flow. The saturations remain approximately constant, although that of the oil is lower upstream due to the vaporization of its most volatile components. The temperature of the matrix is practically equal to that of the steam. In this zone three fluid phases exist, of which only two, the water and gas phases, are flowing.

Zone II. The steam comes into contact with a cooler matrix and condenses, so that although the average temperature (i.e. the only temperature that can be measured in a porous medium, using for example a thermocouple) lies between that of the steam and that of the matrix, an equivalent thermal conductivity can no longer be determined (see Section 61.2). During condensation, the average temperature decreases and the previously vaporized hydrocarbons condense at the same time as the steam.

Zone III. In this zone the displacement is by hot water. However, since the specific volume of steam is far greater than that of water, the velocity of the water is higher than if liquid water had been injected at the same mass flowrate.

The relative importance of vaporization and condensation of the most volatile components of the oil depends on its chemical composition. It may be estimated that the increase in recovery due to thermal expansion varies from 3 to 5% of the oil in place for hot water displacement (lighter oils having higher coefficients of expansion) and that the increase due to steam distillation may reach 20% for

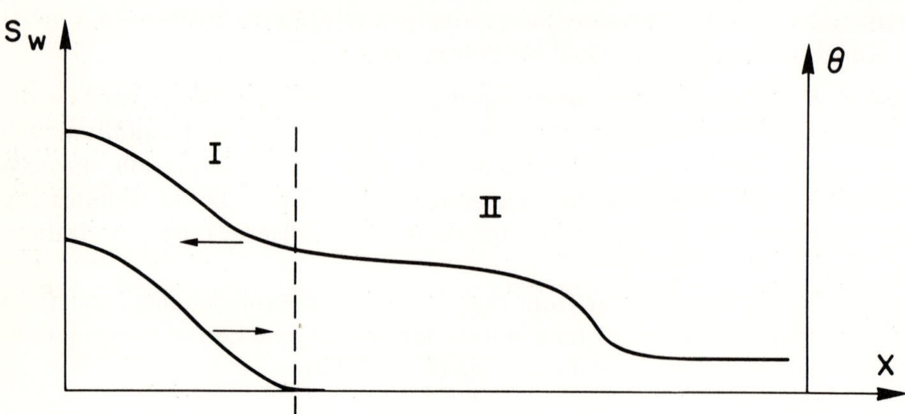

Fig. 62.11.

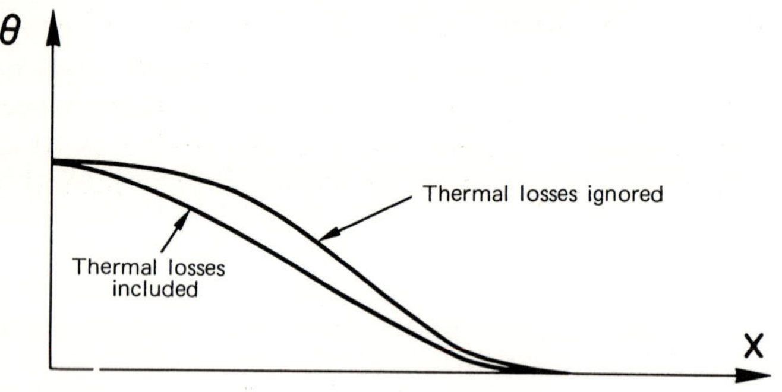

Fig. 62.12.

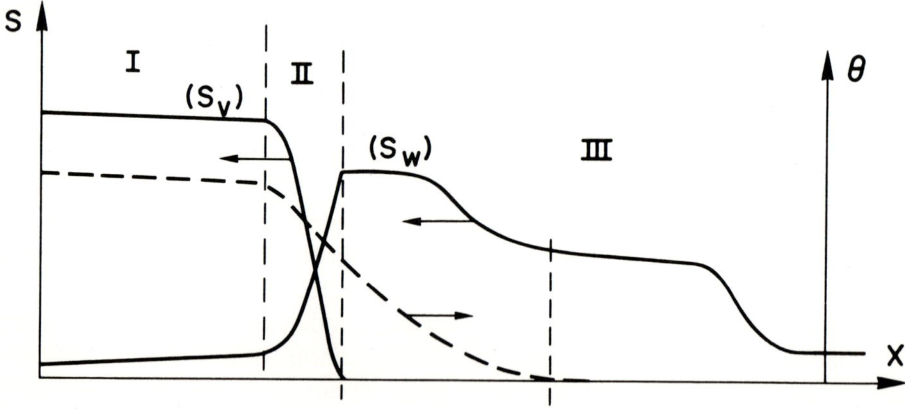

Fig. 62.13.

a crude which is 50% distilled at 170° C. The residual oil saturation after flooding is only slightly reduced by an increase in temperature, thus in the case of light oils steam distillation is the largest contributing factor in the efficiency of hot fluid displacement. For heavy oils the converse is true.

If we now take the heat loss to the surrounding formations into account, we find that the steam condenses continuously upstream and downstream and that the extent of the condensation zone, which starts practically at the injection well, increases considerably with time. Zones I and II can still be distinguished. In zone I the temperature remains practically constant because the steam condensation there compensates for the heat losses. In zone II, where the condensation front moves downstream heating rock and fluids, the steam saturation decreases to zero.

C. A comparison of displacement by cold water, hot water and steam

As we have seen, for both hot water and steam injection, the recovery at breakthrough of the water front is always better than that achieved by cold water injection. In addition, in the case of steam injection, oil is produced even after water breakthrough, until breakthrough of the steam front. This has been confirmed by laboratory experiments on linear systems, the results of which have shown recoveries often higher than 80% at steam breakthrough (Fig. 62.14).

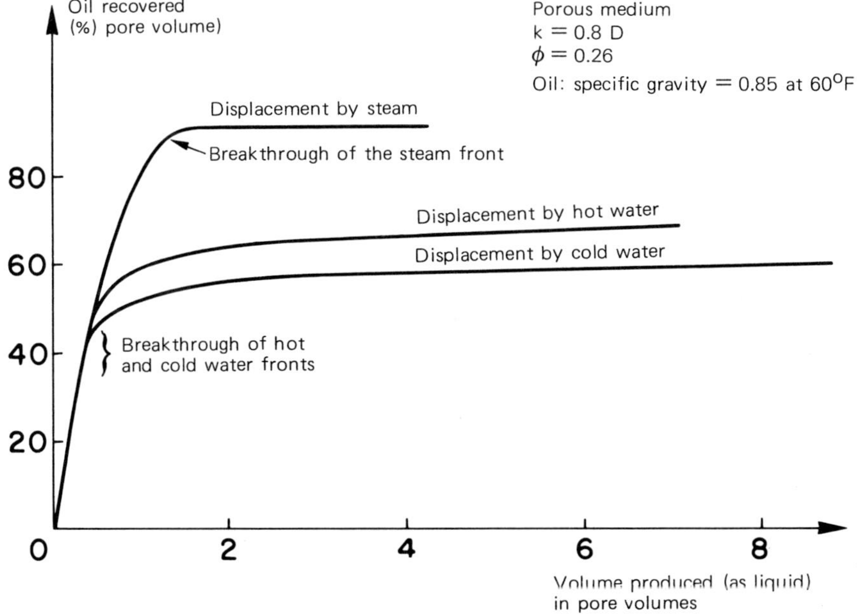

Fig. 62.14.

62.2. Areal sweep efficiency and stability

In Section 15 it was shown that both the area swept during an isothermal displacement and the stability of the displacement are highly dependent on the mobility ratio M. If it is assumed that the relative permeability k_r is independent of viscosity then as a first approximation the areal sweep efficiency and stability depend on the viscosity ratio of oil to water μ_o/μ_w. The greater this ratio, the lower the areal sweep efficiency. To illustrate this, let us use the concept of frontal displacement.

If we consider a hot water displacement with no mass transfer between phases, piston-like from a thermal rather than hydrodynamic point of view, and with a high viscosity ratio μ_o/μ_w (viscosity of oil at reservoir temperature/viscosity of water at injection temperature), the areal sweep efficiency will be low and the displacement may become unstable. This is always the case around an injection well at the start of injection.

In practice, as the front advances, the horizontal thermal conductivity, the thermal capacity of the rock and fluids in place and the forced convection result in the oil being heated and the injected water cooled, even if there is no heat loss to the surrounding formations. This zone of gradual temperature change with distance grows with time, and the assumption of a temperature front becomes invalid.

Additionally, if instability occurs, viscous fingering will carry heat downstream by convection and it will be conducted laterally to oil and rock as yet insufficiently heated. This will extend the temperature profile still further. The same effect applies to the saturation profiles, and the assumption of a hydrodynamic front cannot be maintained.

The result is that after a certain time the oil is displaced by a fluid at the same temperature, thus the areal sweep efficiency is better than that which would be calculated with the assumption of a thermal front.

If we still wish to consider a steam displacement while retaining the notion of thermal and hydrodynamic fronts, a ratio of the form μ_o/μ_w is no longer a useful parameter. At the saturation temperature, the ratio of the specific volumes of steam and water is 86.4 at 20 bar and even at 100 bar is still 12.8, so that throughout the condensation front the fluid velocities cannot be equal. The velocity ratio depends on both the pressure and the thermal capacity of the rock. If the assumptions made at the start of this section are retained the parameter μ_o/μ_v (where the subscript v refers to water vapour or steam) must be replaced by $\mu_o V_o/\mu_v V_v$, where V represents velocity.

Using Fig. 62.21 we can compare, for equal oil displacement velocities, unidirectional displacement by saturated, dry or superheated steam (subscript v), water at saturation temperature (subscript c) and water at reservoir temperature (subscript w) (Ref. 12). The assumptions made are those given at the start of

μV / μV
Hydrostatic pressure
Normal geothermal gradient (3°C/100 m)
Surface temperature 20°C
Tv = Tc saturation
μv Vv / μc Vc
Tv = 350°C
300°C
250°C
200°C
φ=0.1
φ=0.2
φ=0.3
Tv saturation
μv Vv / μw Vw
μc Vc / μw Vw
Tc saturation
10
9
8
7
6
5
4
3
2
1
0.9
0.8
0.7
0.6
0.5
0.4
0.3
0.2
0.1
50
100
200
300
400
600
800
1000
Z (m)

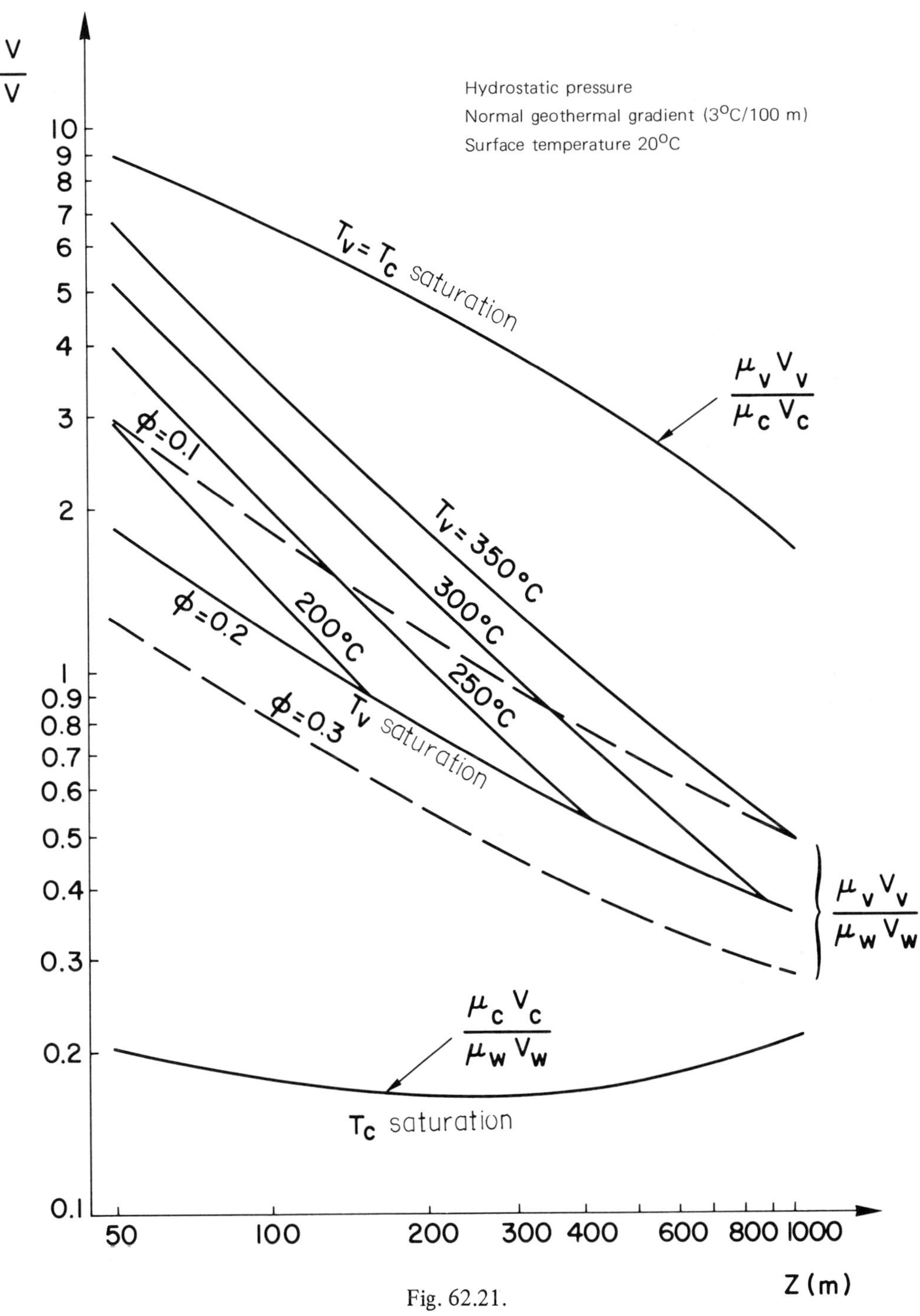

Fig. 62.21.

Section 62. This figure gives the μV ratios to a depth of 1 000 m. As concerns areal sweep efficiency and stability, we can see that:

(a) For a reservoir with 20% porosity, displacement by saturated steam is more efficient than that by water at reservoir temperature, down to around 130 m, below which the latter is more efficient.

(b) Even a relatively low superheat greatly increases the depth down to which steam is more efficient than cold water.

(c) Steam is more efficient at low porosities.

(d) The degree of instability of hot water displacement compared to that of cold water is practically independent of depth.

These results illustrate the general behaviour, and the remarks made earlier for hot water displacement concerning the flattening of the temperature profile with time remain equally valid for steam displacement. Moreover, when fingering occurs the lateral transfer of heat by conduction has a stabilizing influence, since it causes preferential condensation of the steam which has advanced furthest in the porous medium.

These conclusions are equally applicable to "wet combustion" which gives a higher areal sweep efficiency than dry combustion and which will be discussed later.

62.3. Heat loss

Since the heat is carried some distance by the displacing fluid to its final destination in the reservoir, heat loss is a major factor in the use of the various methods of recovery by hot fluid injection.

A. *Heat loss from the reservoir*

Although the heat lost to the surrounding formations does not reduce the velocity of the hot zone it does reduce its temperature, except in the steam condensation zone. However, the extent of the latter zone is reduced and consequently so is the thermal efficiency of the process. It is therefore important to determine the order of magnitude of the heat loss. The simplest method (Ref. 13) is to consider the reservoir to be a piston-like model with the following assumptions (Fig. 62.31):

(a) In the surrounding formations, heat transfer parallel to the bed is neglected, thus only the vertical component of the temperature gradient $\partial\theta/\partial z$ is taken into account.

(b) In the porous medium of thickness Z, the dynamic and thermal processes are unidirectional.

z
A(t1)
A(t2)
Z/2
θ∞
θc
θ
at time t

z
A(t1)
A(t2)
Z/2
Hot fluids
Cold fluids
θ

A(t)
A(t2)
A(t1)

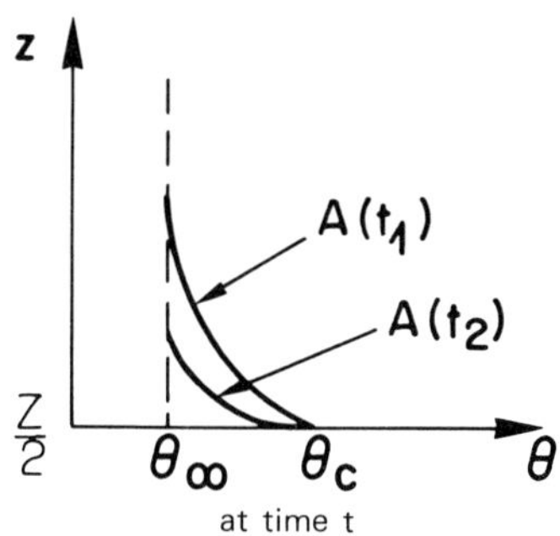

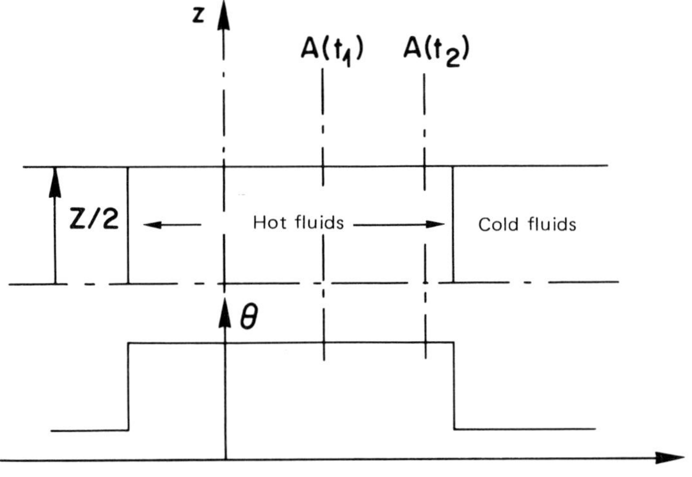

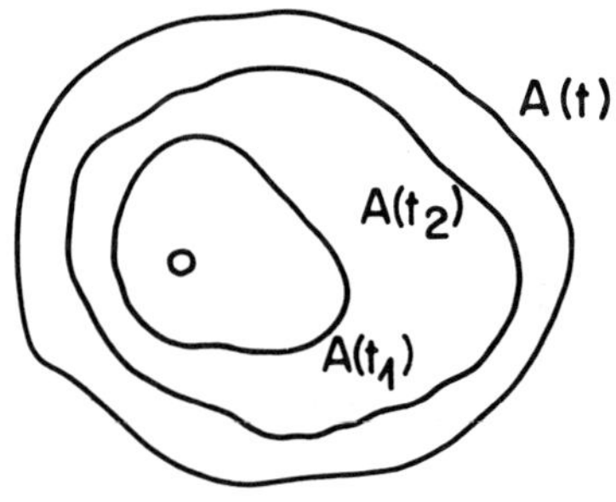

Fig. 62.31.

(c) There is a temperature discontinuity between the constant temperature hot zone and the zone at reservoir temperature, thus longitudinal heat transfer by conduction and convection is neglected.

The initial reservoir temperature is θ_∞. Let $\psi(t)$ be the heat flux corresponding to the heat loss and $A(t)$ the area swept by the hot zone at temperature θ_c; let M be a point in the reservoir reached by the hot zone at time τ.

The heat transfer equation for the surrounding formations is:

$$a_o \frac{\partial^2 \theta}{\partial z^2} = \frac{\partial \theta}{\partial t}$$

where

θ is the temperature and,

$a_o = \dfrac{\lambda_o}{(\rho c)_o}$ is the thermal diffusivity of the medium.

If we only consider the overburden and neglect the geothermal gradient, the following conditions can be set:

$$z > \frac{Z}{2}$$

$$\theta(z, t) = \theta_\infty \quad \text{for } t < \tau \quad \text{(initial condition)}$$

$$\theta\left(\frac{Z}{2}, t\right) = \theta_c \quad \text{for } t \geqslant \tau \quad \text{(boundary condition)}$$

The differential equation has the following solution([1]):

$$\frac{\theta - \theta_\infty}{\theta_c - \theta_\infty} = \operatorname{erfc}\left(\frac{1}{2\sqrt{a_o}} \frac{z - Z/2}{\sqrt{t - \tau}}\right)$$

The heat loss $d\psi(t, \tau)$ to the overburden adjacent to the incremental area dA representing the increase in the extension of the steam zone between time τ and $\tau + d\tau$ is thus:

$$d\psi(t, \tau) = -\lambda_o \left(\frac{\partial \theta}{\partial z}\right)_{z=\frac{Z}{2}} dA = \frac{(\theta_c - \theta_\infty)\lambda_o}{\sqrt{a_o \pi (t - \tau)}} \frac{dA}{d\tau} d\tau$$

([1]) $\operatorname{erfc}(x) = \frac{2}{\sqrt{\pi}} \int_x^\infty \exp(-\alpha^2)\, d\alpha = 1 - \frac{2}{\sqrt{\pi}} \int_o^x \exp(-\alpha^2)\, d\alpha$

The value of this error function can be found from tables.

If we assume that the characteristics of the underburden are identical to those of the overburden, the total heat loss at time t will be given by:

$$\psi(t) = 2\int_o^t d\psi(t,\tau)\,d\tau = \frac{2(\theta_c - \theta_\infty)\lambda_o}{\sqrt{a_o \pi}} \int_o^t \frac{1}{\sqrt{t-\tau}} \frac{dA}{d\tau} d\tau$$

In order to evaluate this integral the areal development of the steam zone must be known:

Let $\psi_i(t)$ be the quantity of recoverable heat injected into the reservoir per unit time and $(\rho c)_c^*$ be the equivalent specific heat of the hot zone. A thermal balance may be written as follows:

$$\psi_i(t) = (\rho c)_c^* Z(\theta_c - \theta_\infty)\frac{dA}{dt} + \psi(t)$$

If $\psi_i(t)$ is a constant ψ_o, an analytic solution may be obtained:

$$A(t) = \frac{\psi_o (\rho c)_c^* Z\, a_o}{4\lambda_o^2 (\theta_c - \theta_\infty)}\left(\exp(\omega^2)\operatorname{erfc}(\omega) + \frac{2\omega}{\sqrt{\pi}} - 1\right)$$

with

$$\omega = \frac{2\lambda_o}{(\rho c)_c^* Z\sqrt{a_o}}\sqrt{t} \text{ (dimensionless)}$$

The advantage of this model is that it makes no assumption as to the shape of the heated surface. If $\psi_i(t)$ is not constant, we may use the same method by dividing $\psi_i(t)$ into a series of steps of constant ψ_i. Unfortunately this model, based on a thermal balance, assumes piston-like displacement of the oil.

Although the results calculated using this model for the extent of the heated zone are far from those obtained experimentally, principally because it assumes that the heated zone is entirely at steam temperature, the thermal efficiency (fraction of injected heat remaining in the reservoir) thus calculated is in fairly good agreement with experimental results.

We have seen then that, as might have been expected, recovery methods involving hot fluid injection are not applicable to very thin beds or to those in which the spacing between injector and producer is large, which is often the case for deep formations.

B. In the injection well

A further cause of heat loss occurs in the passage of the hot fluids in the injection well from surface to the injection zone. Since the various formations encountered are colder than the injected fluid, part of the enthalpy of the fluid is lost, either by decrease of temperature in the case of hot water, or by reduction of quality in the case of steam. Heat transfer takes place by the three classic mechanisms: conduction, convection and radiation. Efficient well completion techniques can reduce the heat losses. Of course, these losses decrease with time, but we can see that they form an additional limit to the exploitation of deep formations by this method.

The heat losses can be evaluated by means of the calculations described in Ref. 14.

62.4 Comparison of hot fluid injection techniques

In spite of its high cost, displacement by steam generally appears to be more favourable than that by hot water because of its greater stability, higher sweep efficiency and the larger quantity of thermal energy carried per unit mass of injected fluid.

However, hot water may become competitive in reservoirs containing relatively light oils, of viscosities of the order of a few hundred centipoise, as long as the well spacing is large or the pressure high, so that a given quantity of injected heat contacts larger surfaces at lower operating temperatures.

If the reservoir has already been waterflooded, hot water injection may again be the most suitable, since a rather large volume of steam would be required to heat and displace such large quantities of water.

We have seen that heat losses play a large part in the thermal efficiency of the operation, which at most can only reach 50%. If, in addition, we take account of the fact that a large part of the heat is retained in the reservoir rock, we see that only a small part of the injected thermal energy is used to heat the fluids.

The problem is therefore to obtain the highest recovery for the lowest expenditure of thermal energy. Several methods have been proposed and applied with this aim:

(a) Cyclic injection of steam (also known as the "huff and puff" or "steam soak" method) followed by a steamflood.

(b) Injection of a steam slug followed by cold water.

(c) Injection of hot water.

A. Cyclic injection of steam possibly followed by a steamflood

The cyclic technique, which requires only one well used alternately as injector and producer, clearly reduces the adverse influence of the mechanisms described on the thermal efficiency of the operation, and enables better utilisation of the energy supplied. Such a process, which allows the immediate surroundings of a production well to be maintained at a higher temperature than that of the formation, is currently a widely used stimulation technique. It has the advantage of improving the flow of oil near the wellbore, where the pressure gradient is normally the largest. Heat is transported by steam injected into the formation.

This technique is preferably applied to reservoirs which still have some natural drive, since most of the mechanical energy carried by the steam is converted into thermal energy (latent heat of condensation). In general one cannot rely on the injected energy to assist the oil backflow, since the volume of water injected, when in liquid form, represents only about 1% of the pore volume involved, and the small increase in pressure near the well is rapidly dissipated. This method may also be applied to reservoirs being pumped.

This technique comprises three phases:

(a) An injection phase comparable to steamflooding.

(b) A soak period during which the well is shut-in. The heat diffuses and all the injected steam condenses, resulting in an increase in oil temperature and a subsequent reduction in warm water production during the third phase. Laboratory experiments indicate that this soak period has very little effect on the production.

(c) A production phase. After recovery of part of the condensed water, oil production stabilizes at a much higher level than that obtained before the start of the process. The heat lost by conduction to the surrounding formations and to the matrix is partly recovered during the production phase, but at a lower temperature, and the productivity continually declines.

The preceding cycle can, of course, be repeated. The lower the normal rate of production due to high oil viscosity, the more economic the process becomes. The increased production rate may justify the drilling of additional wells. If the viscosity of the oil is too high the oil will not be displaced from the cold zone to the hot zone, and only the hot zone will be productive.

When, after many steam stimulation cycles, the reservoir energy becomes too low to support economic production rates, steamflooding from well to well may be considered, taking advantage of both the reduced reservoir pressure, which allows the use of a less expensive steam, and the increased formation temperature due to the earlier stimulations.

B. Injection of a steam slug followed by cold water

As we have seen, steam has the advantage of carrying a very large amount of heat per unit mass; but very often its temperature, imposed by the reservoir pressure, is higher than that required to obtain the desired mobility ratio. Thus after injecting a steam slug of a size determined by the formation characteristics and well spacing, cold water is often injected. This recovers the thermal energy stored in the reservoir rock, and a small part of that contained in the surrounding formations. The temperature of the hot water thus obtained is high enough to ensure a high recovery efficiency.

C. Hot water displacement

As discussed above, it is sometimes preferable to inject hot water, but care should be taken to avoid preferential channelling to which hot water is more susceptible than steam.

62.5. Field applications of hot fluid injection

In view of the application of hot fluid injection, a number of mathematical models have been developed, both analytic and numeric. The data used in these models for the rock and fluid properties are taken from laboratory experiments. Current models only provide estimates for ideal reservoirs, due to the many simplifying assumptions made. They ignore many secondary phenomena which are difficult to quantify and thus cannot give fully reliable forecasts. They will not be discussed further.

We shall limit our investigation to a compilation of the published results of a number of field trials. Moreover, no absolute significance should be attributed to these results. If, for example, we consider cyclic injection, the field tests have been carried out in such varied situations that it is impossible to give a range of conditions which would guarantee the success of such a project. If in addition we note that there is naturally little or no information on failures (and there have certainly been many), which would have given valuable data, we are right to be cautious in defining the limits of application of the various methods, and prefer to show only the apparent trends.

However, a number of general conclusions can be drawn, some of which deal with points which are not considered in model studies. Thus amongst the parameters to be examined before any practical application, the following are of particular importance:

(a) Reservoir parameters which define the limits of the process:
- Permeability.
- Oil content and properties.
- Qualitative effects of hot fluid injection, related to the behaviour of the oil and the matrix.
- Thickness, depth, stratification and heterogeneity of the formation.
- Reservoir pressure.

(b) Operating parameters:
- Injection rate, steam quality.
- In the case of displacement, well spacing.
- In the case of stimulation, injection time, soaking time, production time and production rate.

These parameters are all independent and each has an influence on the efficiency of hot fluid injection. This is directly related to the thermal efficiency of the process, the ratio of energy received by the formation to that originally expended. The best operating method is, of course, that which gives the maximum thermal efficiency.

A. *Limitations of the methods*

1. *Permeability*

For the same mass flowrate, the pressure gradient due to the flow of steam is higher than that for water flow. For example, at 75 bar the water/steam kinematic viscosity ratio is around 4. Thus the injectivity index declines when the quality of the injected steam improves, or as the steam proceeds further into the formation. This means that if the injection pressure is held constant the rate will decline. We could conclude then, that a lower permeability limit must exist, below which the profitability of the process will be doubtful, assuming that secondary phenomena are not important. In the tests made to date, a limit of 300 mD appears to have been the rule. For full-scale displacements, it is preferable to have a permeability no lower than 1 darcy.

2. *Oil content and properties*

There is no technical restraint which determines the minimum oil content required, but the need to operate profitably imposes a practical minimum value. In Californian field tests of steam stimulation a limit of 16% has been observed, but in other tests an oil content as low as 10-12% has been regarded as the minimum. Note that oil content is defined as the volume of oil contained in unit volume of the porous medium.

The viscosity of very heavy oils is extremely high and it is important to reduce it as far as possible. However, the viscosity reductions obtained are not always sufficient to make the oil particularly mobile. In practice, reservoirs which can be successfully exploited are those containing oil of medium viscosity (from 8 000 to 50 cP approximately). It has been shown that the methods would give good results with very light oils.

3. *Qualitative effects of hot fluid injection*

An increase in the temperature of the solid matrix and the passage of steam and its subsequent condensation within the pores may have certain secondary effects which should sometimes be taken into account:

(a) Rock wettability changes due to the presence of steam: although this effect has been observed in some pilot tests it is rarely a cause of failure.

(b) Swelling of certain clays by the fresh condensed water: clay stabilization by the injection of a solution containing a tri-valent cation has been successfully tested.

(c) Formation of emulsions.

(d) Clean-up effect: this favourable effect has been noted. The porous matrix is often "damaged" in the vicinity of the wellbore, being partially blocked by heavy oil fractions deposited during primary production. Thus a brief period of steam injection may be used prior to water injection in order to improve the injectivity of the well.

4. Thickness, depth, stratification and heterogeneity of the formation

In the selection of a reservoir for hot fluid injection, there are two principal parameters to be considered:

(a) The relative amount of heat losses, which depends on both the thickness and depth of the reservoir and,

(b) The technical aspects of high pressure injection, which place a limit on reservoir depth.

It is generally considered that the formation thickness must be greater than 10 m and that its depth should not be greater than 1 000 m.

In the case of well to well displacement however, the performance in thicker formations may be hampered by gravity segregation effects, since the steam will travel preferentially along the top of the bed. In a stratified reservoir, the steam will enter the most permeable strata preferentially, and this may be one of lower layers. Oil resaturation of this layer then occurs by segregation. This process, originally proposed for the recovery of bitumen, has been applied in the Slocum oilfield in which a water-bearing strata underlies the oil-bearing strata. Considerable oil production was obtained, but at the expense of high steam consumption (Ref. 15).

The fact that a reservoir is stratified or even that it contains random heterogeneities often leads to the preferential channelling of the steam. Early breakthrough will then occur at the production wells, which will have to be shut in or selectively plugged, at the expense of oil production.

5. Reservoir pressure

We have seen that when the reservoir pressure is insufficient, steam stimulation is uneconomic, in fact impossible. However, we shall see that by suitable adjustment of the length of the injection and soaking periods it is in certain cases possible to obtain significant oil production.

B. Operating parameters

In the case of hot water or steam displacement, the well spacing is often smaller than for more conventional methods of exploitation in order to reduce the relative importance of heat loss. Furthermore the wells are often arranged so as to minimize the detrimental effects of channelling (line drive or hexagonal or octagonal patterns).

As an example, we give here the typical operating parameters of a steam stimulation trial: the heat injected into the formation per cycle is between 50,000 and 150,000 thermies per metre of thickness, corresponding to an injection period of from 5 to 20 days. The soaking period is several days long and the production may last several months. The second cycle commences when the well productivity becomes relatively stable following its decline.

However, this is not the only way in which such processes are operated. For example, in the Kern River reservoir (Ref. 16) containing an oil with a viscosity of 5 000 cP at a pressure of 8 bar, the natural reservoir energy is very low. Maximum use is therefore made of the energy available due to the increase in pressure near the well during injection (a result of the high initial viscosity of the oil), by putting the well back on production as soon as the injection ends. This is achieved by injecting the steam down the annulus and pumping the well immediately injection stops, the pump being already in place. The heat losses involved in this operation are high, but it is estimated that they are largely compensated by the resulting increase in oil production. The short production cycles used, being around 4 months each, confirm the temporary nature of the pressure increase produced by steam injection.

C. Test results

During well to well steam displacement for oils of viscosities in the range 1 000 to 5 000 cP, it is estimated that the areal sweep efficiency will be at least 50%, sometimes reaching as high as 90%. Vertically the steam generally occupies no more than 40% of the reservoir thickness. Oil recovery varies from 35 to 50% of the oil in place in the part of the reservoir considered. In the zone swept by steam the residual oil saturation lies between 8 and 22%, compared with 35 to 45% for the zone swept by hot water. A comparison of recovery after displacement by cold water, hot water and steam, assuming the same reservoir and the same well configuration, was shown in Fig. 62.14.

In steam stimulation trials, the production achieved depends principally on the available reservoir energy, its gradual decompression making successive stimulation cycles less productive. The principal result of steam stimulation is an acceleration in production which makes the wells more economically attractive. The recovery factor is lower than that which would be obtained by steam displacement, but it often exceeds the recovery which would be obtained by one of the classical methods of exploitation.

D. Field examples

(a) The Schoonbeek field in Holland is produced by displacement by hot brine and steam. The oil is of medium viscosity, 175 cP at most (Ref. 17).

(b) In California and Venezuela, more than 1 000 wells produce by cyclic steam injection.

(c) The East Coalinga field in the USA is produced by steam injection followed by cold water injection (Ref. 18).

63. IN-SITU COMBUSTION

The principle of in-situ combustion is to achieve combustion within the pores of a hydrocarbon-bearing reservoir, burning part of the oil in place in order to improve the flow of the unburned part. Combustion is supported by the injection of air into the reservoir at one or more wells. The heat generated during combustion is sufficient to raise the rock to a high enough temperature to enable the combustion front to self-propagate after initial ignition.

63.1. Methods of in-situ combustion

If we consider the simple case of one air injection well and one production well, it can easily be appreciated that the direction of propagation of the combustion front depends on where ignition takes place.

If the zone surrounding the injection well-bore is brought to a sufficiently high temperature, ignition takes place in the vicinity and the combustion front travels in the direction of the production well, i.e. in the direction of the fluid flow. This is known as forward combustion.

On the other hand, if the zone surrounding the production well is heated, ignition takes place near this well and the combustion front travels towards the injection well, i.e. in the opposite direction to the fluid flow. This is known as reverse combustion.

A. Dry forward combustion

In this process, the combustion front acts as a piston which pushes ahead of it the unburnt crude fractions from the swept zones. The heavy fractions, converted into coke, are burnt with the oxygen in the air injected to sustain combustion.

Under steady-state conditions, the reservoir can be divided into four principal zones, numbered from upstream, as shown in Fig. 63.11:

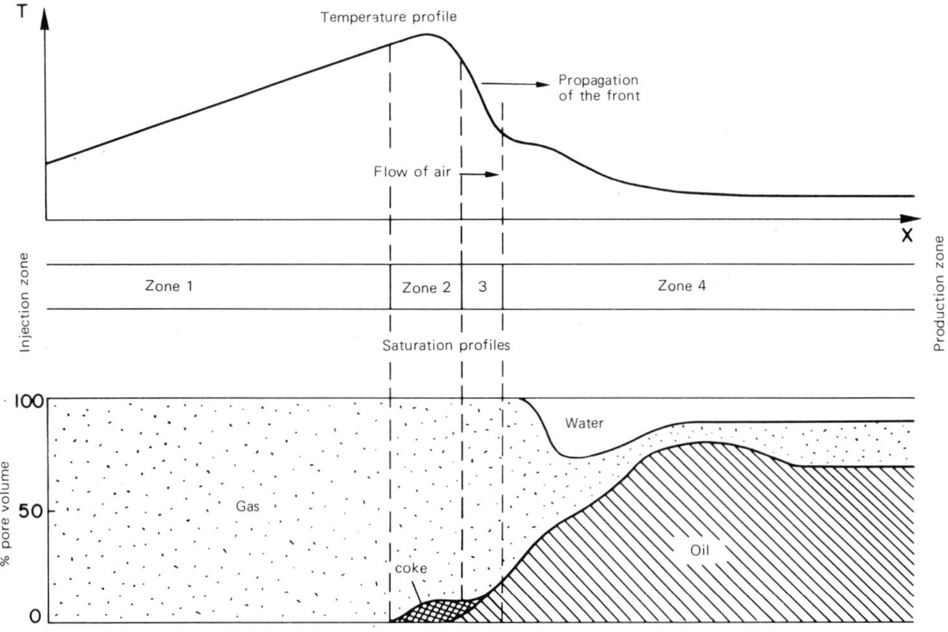

Fig. 63.11. Forward combustion.

Zone 1: Combustion has already taken place and the formation in this zone is completely clean. The injected air is heated by the hot matrix and part of the combustion energy is recovered in this way; temperature decreases upstream.

Zone 2: The combustion zone. Oxygen is consumed by combustion reactions involving the hydrocarbons and the coke remaining on the rock surface. The temperature reached in this zone depends essentially on the nature of the solids, liquids and gases present per unit formation volume.

Zone 3: The coke formation zone. The heavy oil fractions which have been neither displaced nor vaporized undergo pyrolysis. These cracking reactions may occur in the presence of oxygen, if the latter has not been completely consumed in the combustion zone.

Zone 4: Where the temperature has fallen sufficiently, there are no further significant chemical changes. This zone is swept by the combustion gases and displaced fluids, and the following phenomena take place:

(a) In the downstream region nearest the reaction zone, successive vaporization and condensation of the light oil fractions and the interstitial water take place, as does condensation of the water of combustion. This tends to accelerate the downstream heat transfer.

(b) In the region where the temperature is lower than that of water condensation, a zone with a water saturation higher than the initial water saturation exists (water bank) which pushes a zone with an oil saturation higher than original (oil bank). If the oil is highly viscous this may result in plugging of the formation. In every case, these two banks are a zone of high pressure loss. Beyond the oil bank the formation progressively approaches its original conditions.

B. Forward combustion combined with water injection ("wet combustion")

During forward combustion, the enthalpy stored in the matrix is only partly used to preheat the air injected, the rest being lost to the surrounding formations.

Thus the idea arose of combining water and air injection after start-up of the combustion process. In this way the high thermal capacity of water may be used to recover the enthalpy remaining behind the front and transport it downstream.

The process may be conveniently divided into five zones, as shown in Fig. 63.12:

Zone 1: This zone has already been swept by the combustion front and contains little or no hydrocarbon. However, since the temperature is lower than the boiling point of water, the pores contain a liquid water saturation, the remainder of the space being occupied by the injected air.

Zone 2: Water is in the vapour phase in this zone, and the pores are saturated with a mixture of injected air and steam. The injected water vaporization front is at the boundary between zones 1 and 2.

Zone 3: The combustion zone. Oxygen is consumed in the combustion of the hydrocarbons and of the deposited coke formed in the downstream part of the zone.

Zone 4: The vaporization-condensation zone. The temperature in this zone is close to that of the vaporization of water. Progressive condensation of steam and combustion water takes place in this zone. In addition, some light and intermediate oil fractions are vaporized and carried downstream. If the temperature is high enough certain chemical reactions may occur in this zone.

Zone 5: Just downstream of the vaporization-condensation zone is a zone of high back-pressure, due to the formation of a water bank preceded by an oil bank. Further downstream the formation gradually approaches its initial conditions.

C. Reverse combustion

After ignition, the combustion front moves from the producing well towards the injection well.

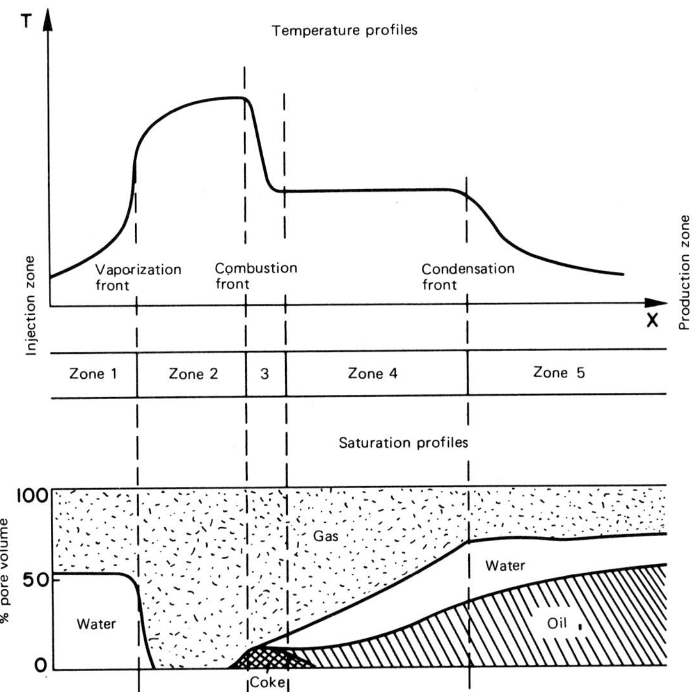

Fig. 63.12. Wet combustion.

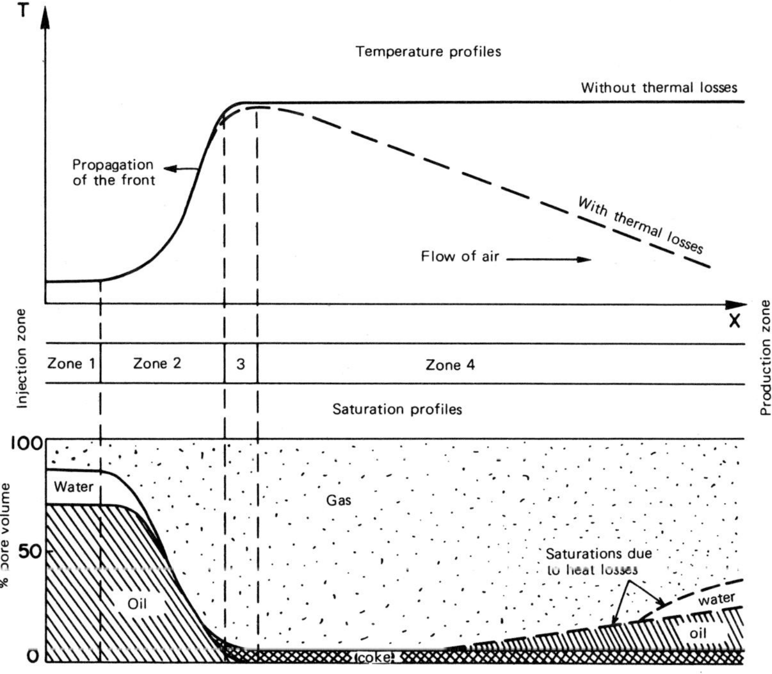

Fig. 63.13. Reverse combustion.

Four zones can be defined, starting from the injection well, as shown in Fig. 63.13:

Zone 1: The formation is at original conditions. However, it is being swept by air, and if the formation temperature and oxidability of the oil are high certain oxidation reactions may occur.

Zone 2: The temperature increases by conduction from the hot zone downstream. The start of oxidation also contributes to the temperature increase. The following phenomena occur: vaporization of the formation water, distillation of the light fractions of the oil and cracking of certain hydrocarbons in the presence of oxygen. The liquid and vapour fractions are displaced downstream, while certain components form the carbon deposit or "coke".

Zone 3: The combustion zone. The temperature reaches its maximum value. The oxidation and combustion reactions involving the most reactive hydrocarbon molecules consume all the oxygen not used by the reactions in the preceding zones.

Zone 4: The unburnt coke remains deposited on the matrix while the vapour and liquid phases flow downstream. If there were no heat losses, the downstream temperature would remain equal to that of the combustion front. In reality, the temperature decreases with distance from the combustion zone. Thus condensation of the distilled oil fractions occurs, and possibly of the steam.

63.2. Laboratory studies of in-situ combustion

A. Crude oil oxidation in porous media

In order to study the reactions which occur during underground combustion, laboratory experiments are performed using core samples impregnated with oil. Air is injected into the cores and their temperature is increased with time so as to simulate the approach of a combustion front (Refs. 9, 19 and 20).

The apparatus used by the *Institut Français du Pétrole* is shown in Fig. 63.21. The sample can be heated to 500° C with the temperature increasing linearly with time (50 – 100° C/hr) as in the case of differential thermal analysis. Thermocouples within the sample allow the thermal effects of the chemical reactions to be observed, and the progress of the reactions is monitored accurately by the continuous analysis of the produced gases.

An example of the results obtained during a typical study is shown in Fig. 63.22([1]). In this case the sample was heated at a rate of 100° C/hr.

([1]) In this chapter, flux and flowrate are given at standard conditions (1 atm, 15° C or 60° F). Standard volumes are shown as sm^3.

Thermocouples
to recorder
Pressure gauge
Air
(or Nitrogen)
Rupture disc
Measurement thermocouples
Oven
Control thermocouple
Insulation
Cell
Control
thermocouple
Seals
Buffer volume
Rotameter
(flow meter)
Oven power
supply
Back-pressure valves
Programmed
Control System
Trap (– 20°C)
To recorder
CO
Analyzer
CO_2
Analyzer
O_2
Analyzer
Gas outlet
Cell
Sintered glass
plate

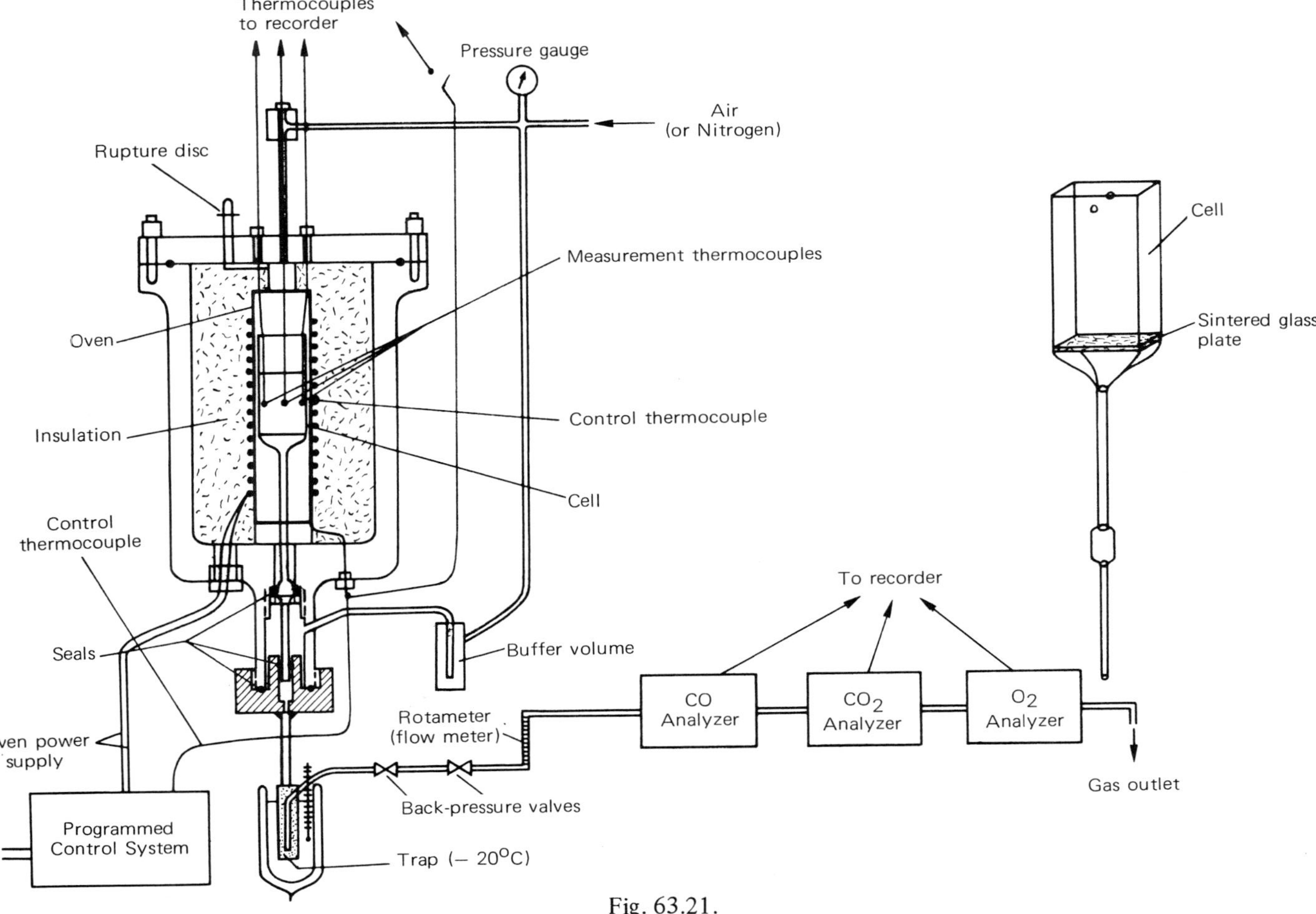

Fig. 63.21.

Temperature °C
500
400
300
200
100
Gas, % vol.
6
5
4
3
2
1
Clean sand
Oil specific gravity: 0.89
Air: 40 sm³/m²hr
p = 5.5 atm.
O₂ consumed
CO₂ produced
CO produced
Sample temperature
Oven temperature
1
2
3
4
5
t, hours

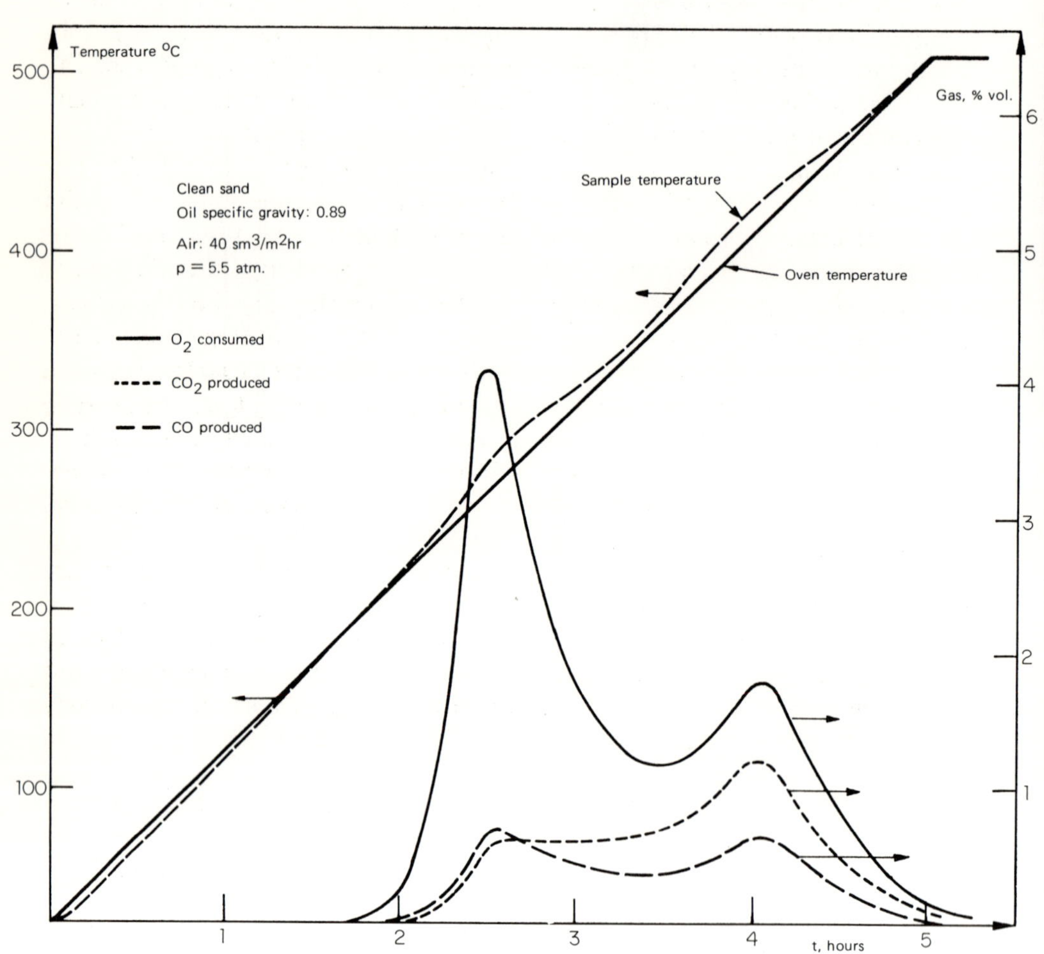

Fig. 63.22.

Oxygen starts to be consumed at between 100 and 150° C. At a slightly higher temperature carbon monoxide and dioxide appear in the produced gas stream.

Two successive reactions may be clearly observed. The first is that of incomplete oxidation of the oil at low temperature ($T < 300°$ C approx.): only a small part of the oxygen consumed is found in the CO and CO_2 produced. The second reaction is that of the combustion of a residue of low hydrogen content: almost all of the oxygen used is found in the CO and CO_2 produced.

The significance of the oxidation reactions depends on the characteristics of both the oil and the matrix (Ref. 9). The two reactions are enhanced by the presence of catalysts based on derivatives of certain transition metals (Cu, Ni, etc.). The oxidability is also increased if the matrix has a high specific surface area (especially in clayey sands) (Ref. 9). These observations lead to the conclusion that at least some of the chemical reactions occur in heterogeneous phase.

Of course, a real in-situ combustion process is far more complex than the process simulated in the laboratory. In particular, even though it may be possible to simulate a temperature profile close to reality, it is practically impossible to impose a realistic injected gas composition profile throughout the sample.

However, these laboratory tests provide extremely useful semi-quantitative data.

For example, during forward combusion the pores containing virgin oil are heated in the presence of a flow of gas with little or no oxygen (Zones 3 and 4 of Fig. 63.11). The first temperature peak is therefore small due to a lack of oxygenated reactant. Whereas, the size of the second peak indicates the amount of coke deposited on the rock surface and burned at high temperature according to the following general reaction:

$$CH_x + \left[\frac{2+\beta}{2(1+\beta)} + \frac{x}{4}\right] O_2 \rightarrow \frac{1}{1+\beta} CO_2 + \frac{\beta}{1+\beta} CO + \frac{x}{2} H_2O \qquad \text{(Eq. 63.21)}$$

The hydrogen/carbon atomic ratio of the coke, x, usually lies between 0.5 and 1.5.

Some laboratory simulations of the approach of a combustion front have been specifically performed to determine the quantity of coke formed and available for forward combustion (Ref. 19), (Fig. 63.23). Experiments were performed on core samples from a particular sand formation saturated with crude oils of various types. The results obtained for the mass of carbon Y_C in the coke per 100 g of the matrix are shown in Fig. 63.23; the oils are characterized either by their density or by the amount of residue formed during a standard test ("Conradson" carbon). It can be seen that the fuel availability tends to increase

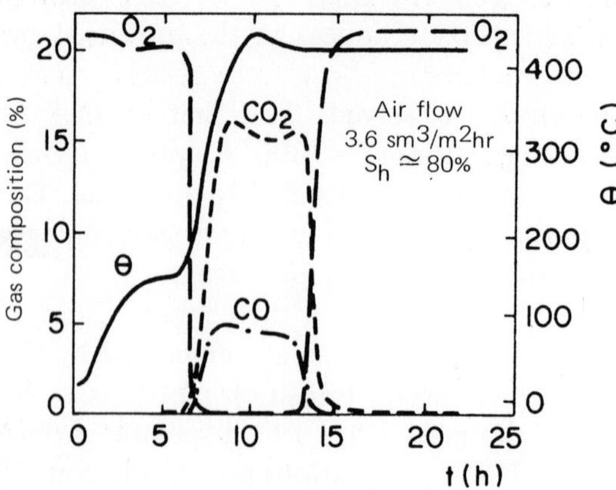

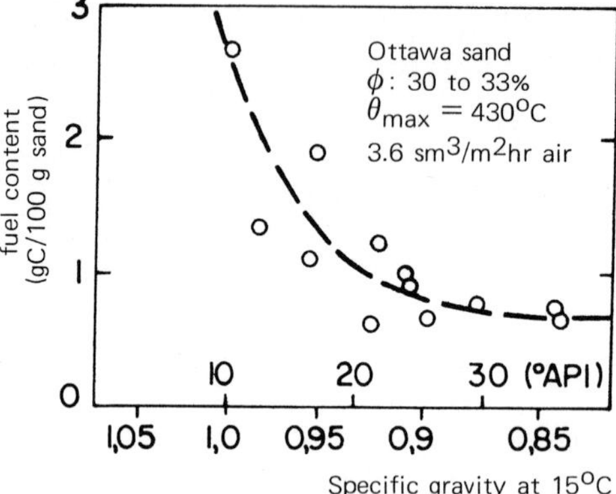

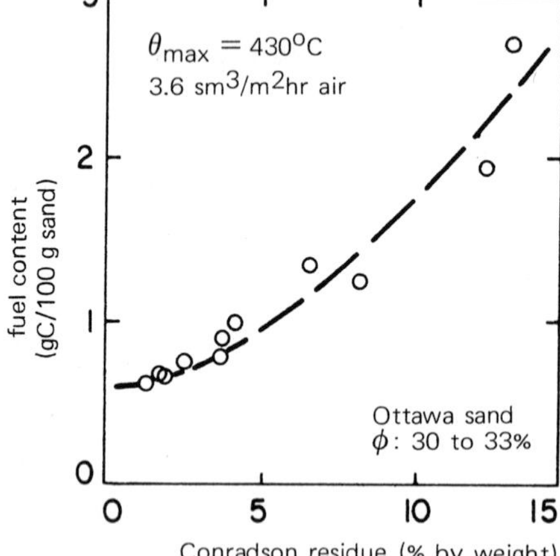

Fig. 63.23. (Ref. 19).

with oil density[1] and with Conradson carbon; however, the dispersion of the data shows that even for an identical matrix there is no unique relationship between these parameters and the fuel availability. Nevertheless, these figures allow an order of magnitude estimate of the amount of coke deposited to be made.

B. Experimental study of forward combustion

1. Apparatus

A schematic drawing of the equipment used for a laboratory study of in-situ combustion is shown in Fig. 63.24. To avoid gravity segregation effects, the horizontal pressure shell is supported on steel rollers that enable the cell to rotate around its axis (alternate rotation at 1 RPM). The thin-walled (2 mm) combustion tube is fitted inside the pressure jacket. The annulus between the two tubes is filled with nitrogen at a pressure equal to the injection pressure. The combustion tube, 2.1 m long and 20 cm in diameter, is surrounded with 17 heating collars which may be used to compensate for heat losses. Ignition takes place by way of a resistance element covering one end of the combustion tube. Thermocouples placed along the axis and at the wall of the combustion tube allow the propagation of the combustion front to be monitored. The produced gases are continuously analyzed and gases may be sampled in situ for subsequent chromatographic analysis. The volumes of oil and water produced are measured and their principal properties determined (density, viscosity, acid index, etc.).

2. Results of dry combustion tests

The effects of operating conditions on the velocity and temperature of the front for a given oil are shown in Fig. 63.25.

It can be seen that the temperature of the front is relatively independent of pressure and air flux density (m^3/m^2 hr). The velocity of the combustion front is roughly proportional to the air flux density but is independent of pressure. This approximation may be expressed as:

$$\frac{\text{air flux density}}{\text{velocity of the combustion front}} \approx \text{Constant} = a \quad \text{(whatever the pressure or air flux)} \qquad \text{(Eq. 63.22)}$$

---

[1] The fuel availability increases when the API-gravity decreases.

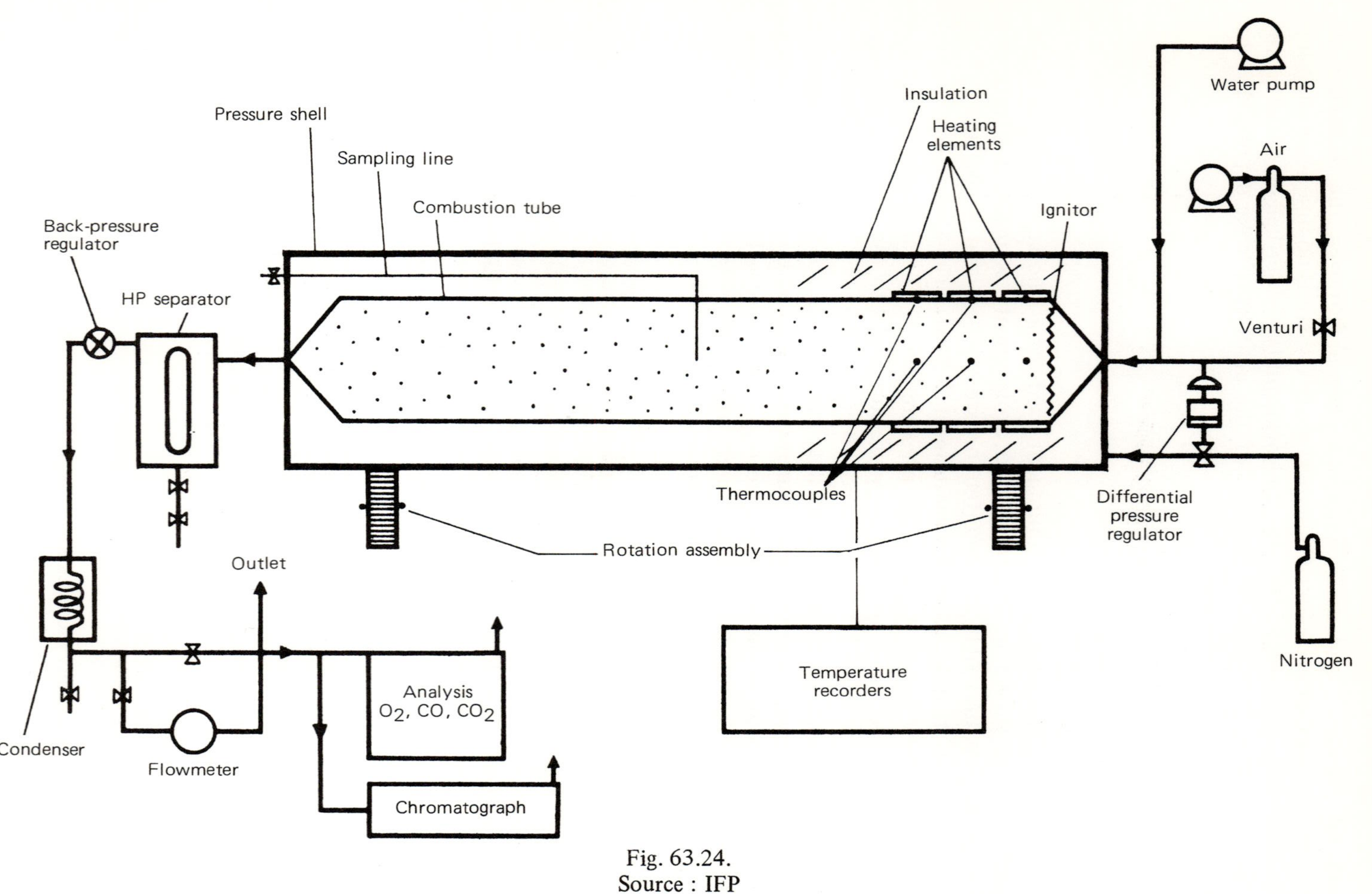

Fig. 63.24.
Source : IFP

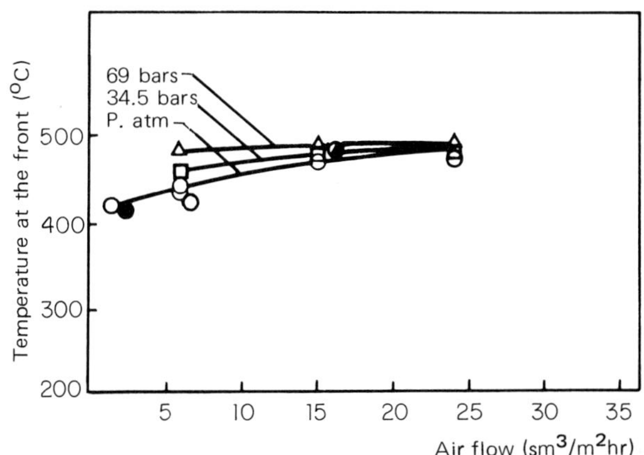

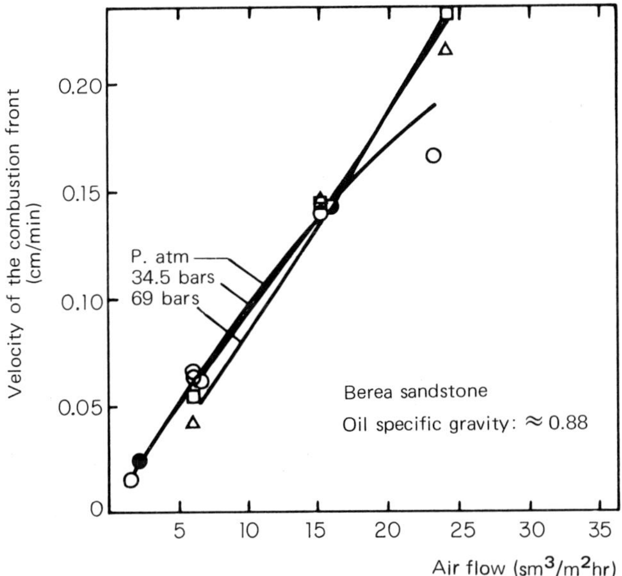

Fig. 63.25.
(Ref. 22).

The parameter a represents the volume of air required to sweep a given volume of the porous medium: it is the ratio of the volume of air per unit cross-sectional area of the combustion front per unit time to the distance travelled by the front in unit time.

In fact, Eq. 63.22 is only valid when the variations of pressure and of air flux density are not too large. For example, in Fig. 63.25 it can be seen that the relationship does not hold for air flux densities greater than 15 sm^3/m^2 hr. This arises because at high air flowrates it is no longer possible to obtain complete consumption of the injected oxygen, especially if the operating pressure is low. Thus the air requirement tends to increase.

In certain porous media the parameter a tends to increase with pressure when operating at low flowrates.

With these restrictions, as a first approximation it can be considered that the amount of air required for dry combustion depends only on the particular oil-matrix combination. The parameter a is dimensionless, being expressed in standard cubic metres of air per cubic metre of rock.

The value of a is normally determined from laboratory experiments on the propagation of combustion fronts. However, it may also be estimated from the fuel availability for forward combustion (Section 63.21). It can easily be shown that the mass of burnt coke per cubic metre of the porous medium, m_b, is related to the percentage of carbon available, Y_C, by the equation:

$$m_b = 10\, Y_C \left(1 + \frac{x}{12}\right) \rho_s (1 - \phi) \text{ kg/m}^3 \qquad \text{(Eq. 63.23)}$$

where

x is the atomic hydrogen/carbon ratio of the fuel (see Eq. 63.21),
ρ_s is the density of the rock (g/cm^3) and,
ϕ is the porosity of the medium.

The air requirement a is related to m_b by a stœchiometric constant: generally about 11 sm^3 of air are required to burn 1 kg of coke. The temperature of the combustion front generally lies between 400 and 700° C, increasing with the fuel availability and the air requirement.

The values of these last two parameters are principally dependent on the oil characteristics (and the effects of the matrix). In the case of light oils they are often low, and for some light oils the fuel availability is insufficient to ensure the self propagation of a forward combustion front. If, on the other hand, a considerable amount of coke is formed, a great amount of air will be required and thus the cost of injection will be high.

As an order of magnitude, an air requirement of between 200 and 400 sm^3 per cubic metre of rock is considered to be reasonable.

During forward combustion, all the unburnt oil is recovered: if the initial oil saturation is high enough, recovery factors of 80-90 % may be achieved in one-dimensional laboratory tests. The produced oil differs only slightly from the oil originally in place, since most of the production is driven ahead of the combustion front and suffers little alteration. However, the specific gravity and viscosity of the produced oil tend to decrease and, at the end of the project, the presence of oxygenated compounds can be detected by the increased acid index (Ref. 25). If combustion takes place efficiently the produced gas will contain little or no oxygen, 12-15% CO_2 and 0.1-4% CO.

3. Results of wet combustion tests

The effect of water injection on the temperature profile has been studied experimentally. Figure 63.26 shows the lengthening of the steam plateau and the reduction of the extent of the high temperature zone (Refs. 23, 24 and 25).

The hydrodynamic effect of water injection is not simply that the condensation front advances faster than the combustion front: the combustion front itself propagates at a higher velocity (Fig. 63.27). This results from an improvement in the efficiency of oil displacement in the steam plateau, and from a corresponding reduction in the amount of coke deposited (Ref. 24).

Highest efficiency for a wet combustion process is achieved when the water/air ratio in the injected fluids is in the range 0.001-0.004 m^3/sm^3. If the water/air ratio is too low, all the injected water will remain trapped upstream of the hot zone and will not assist in transporting energy downstream (Ref. 24). Conversely, if the water/air ratio is too high, the vaporization front will overrun the combustion front and the high temperature zone will disappear.

Oil production is accelerated by the injection of water and the recovery achieved by wet combustion may be even higher than that achieved by dry combustion (Ref. 24).

C. The study of reverse combustion

This method of oil recovery has been studied in the laboratory under a wide range of experimental conditions (Refs. 21 and 22). It appears that the method generally gives lower recovery than the forward combustion methods, due partly to the coke formed remaining unburnt and the combustion of intermediate fractions of the oil, and partly to the recondensation of displaced oil downstream of the combustion front. For these reasons, this method appears to be applicable in only a very small number of cases.

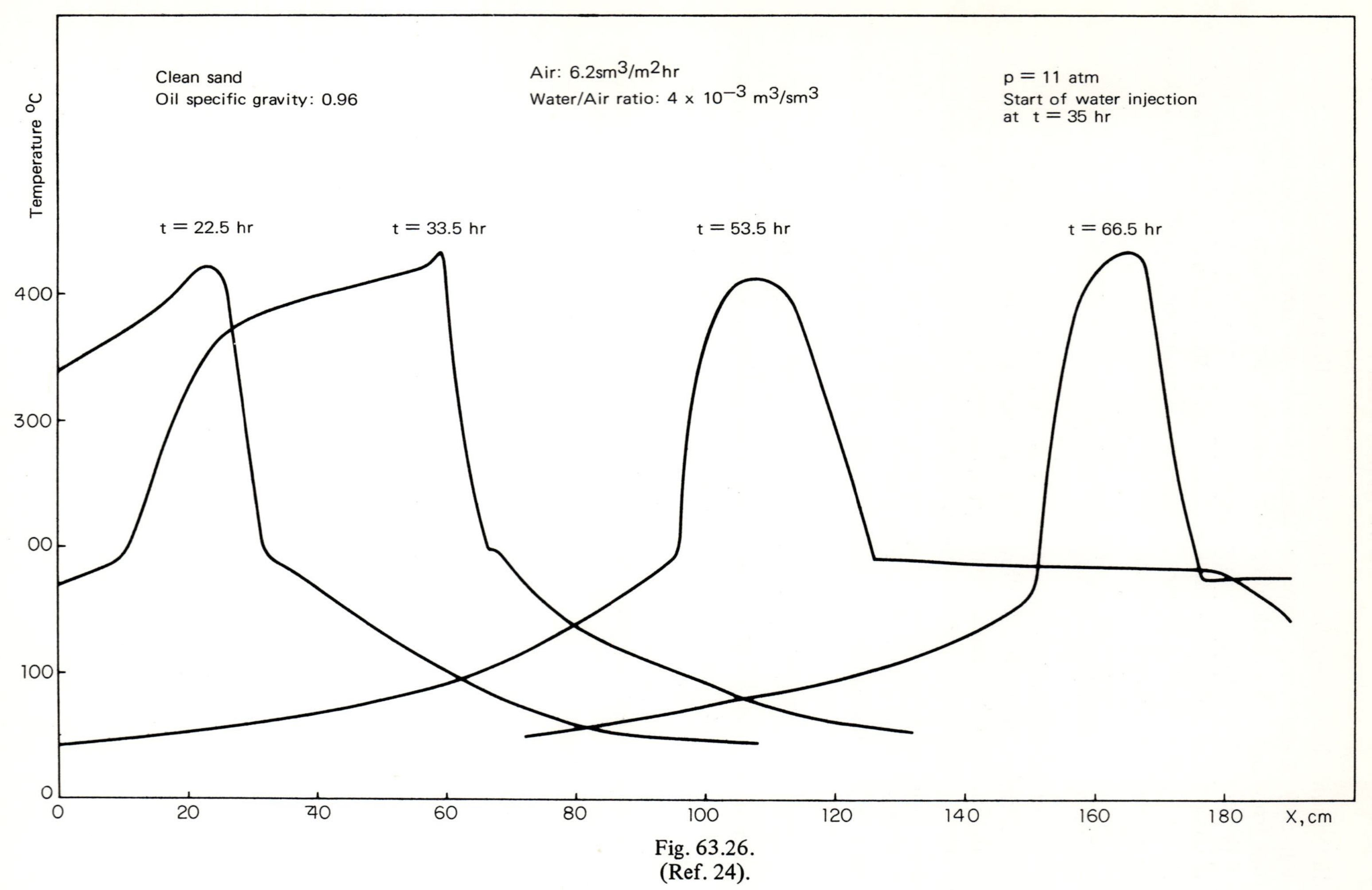

Fig. 63.26.
(Ref. 24).

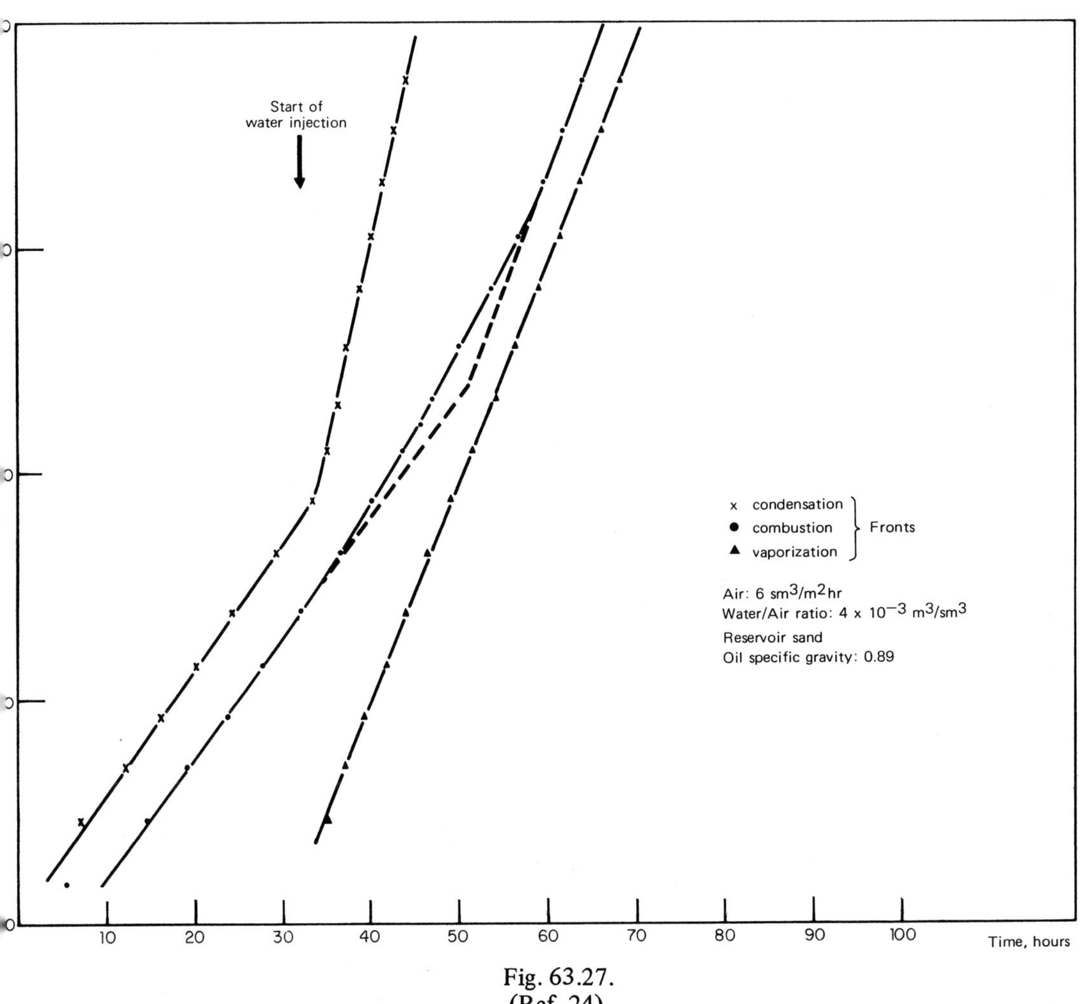

Fig. 63.27.
(Ref. 24).

63.3. Field application of in-situ combustion

A. Advantages of the technique and its limits of application

Compared to the other thermal recovery methods, in-situ combustion has the advantage that the heat is only released in the porous medium itself: there is thus no problem of heat loss in the injection wells. Favourable economic and technical conditions may be found to enable commercial use of the forward combustion method (with or without water injection). This process has several specific advantages: combustion of a heavy residue of low value; production of an oil lighter than that in place and generally not particularly corrosive; solution of CO_2 in the oil resulting in a lowering of its viscosity.

The technical limitations to the application of in-situ combustion methods are principally:

(a) Depth of the formation: limited to around 1 500-1 800 m due to the high air injection pressure which would be required in deeper formations. Neither can this technique be used if the depth of the reservoir is less than 50 m (the overburden may not be able to withstand the injection pressure).

(b) Thickness of the formation: must be thicker than about 2 m (to avoid excessive heat losses).

(c) Permeability: preferably at least 100 millidarcies.

(d) Oil and matrix characteristics: the oil should not be too viscous, to prevent formation plugging by the oil bank, and sufficient coke must be deposited to feed the combustion process. It is often considered that these conditions may be met if the oil specific gravity is between 0.8 and 1.0 g/cm^3. However, this specific gravity range is merely indicative, due to the unknown interaction between oil and matrix and the subsequent effects on both flowing properties and chemical reactions.

In addition to these technical limitations, economic factors must be considered. In particular, it is important that the air requirement is not too high, since air compression is an expensive part of the in-situ combustion process.

However, the most significant economic factor is the ratio of the amount of air injected to that of oil produced: under optimum conditions it lies between 1 000 and 2 000 sm^3 air/m^3 oil. For this ratio to be economically acceptable the initial oil saturation should be large compared with the volume of coke burned to feed the combustion (the latter being of the order of 10 to 40kg/m^3 of formation). It is often considered that the limiting initial oil content for the application of in-situ combustion is of the order of 100 litres of oil per cubic metre of formation.

Wet combustion appears to be more economically attractive than dry combustion: the injection of water, at relatively low cost, results in a large reduction in the volume of air required to recover a given volume of oil (Ref. 23). The use of high water/air ratios is in some cases very attractive (Refs. 23 and 25): under certain conditions oxidation at the temperature of vaporization of water may take place, and may propagate through the porous medium even after the high temperature zone has been extinguished (Ref. 23). However, this particular method of in-situ combustion, known as "superwet" combustion, has not yet been fully studied.

B. Operating methods

Applying a combustion process in the field is obviously rather more complex than it is in the laboratory, where only one-dimensional processes are studied. The shape of a combustion front is affected by the geometrical distribution of the injection and production wells, by the heat losses to the surrounding formations, by reservoir heterogeneities and by gravity. The latter is particularly important in thick formations, in which the injected air tends to channel preferentially to the top of the formation.

1. Technical problems of in-situ combustion

(a) Air compression: multi-stage reciprocating compressors are used, the compression ratio per stage being between 2 and 3. The injection pressure is determined by the characteristics of the formation, and the capacity of the compressors by the pore volume to be swept.

For any given injection pressure, the compression ratio for the selected number of stages and the horsepower required per stage can be determined from Fig. 63.31 for a range of overall compression efficiencies (generally about 70% of that of perfect adiabatic compression). An example of the results obtained is given below (Table 63.31).

TABLE 63.31
INJECTION PRESSURE 45 ATM
OVERALL EFFICIENCY 70%

No. of stages	Compression ratio	Horsepower per stage per 1 000 s m^3 air/day	Total horsepower per 1 000 s m^3 air/day
3	3.56	3.39	10.17
4	2.59	2.42	9.68

In this case 4 stages of compression would be chosen.

For wet combustion the injection wells must also be provided with water injection pumps. The air and water may either be injected simultaneously as a mixture, or alternately in successive slugs.

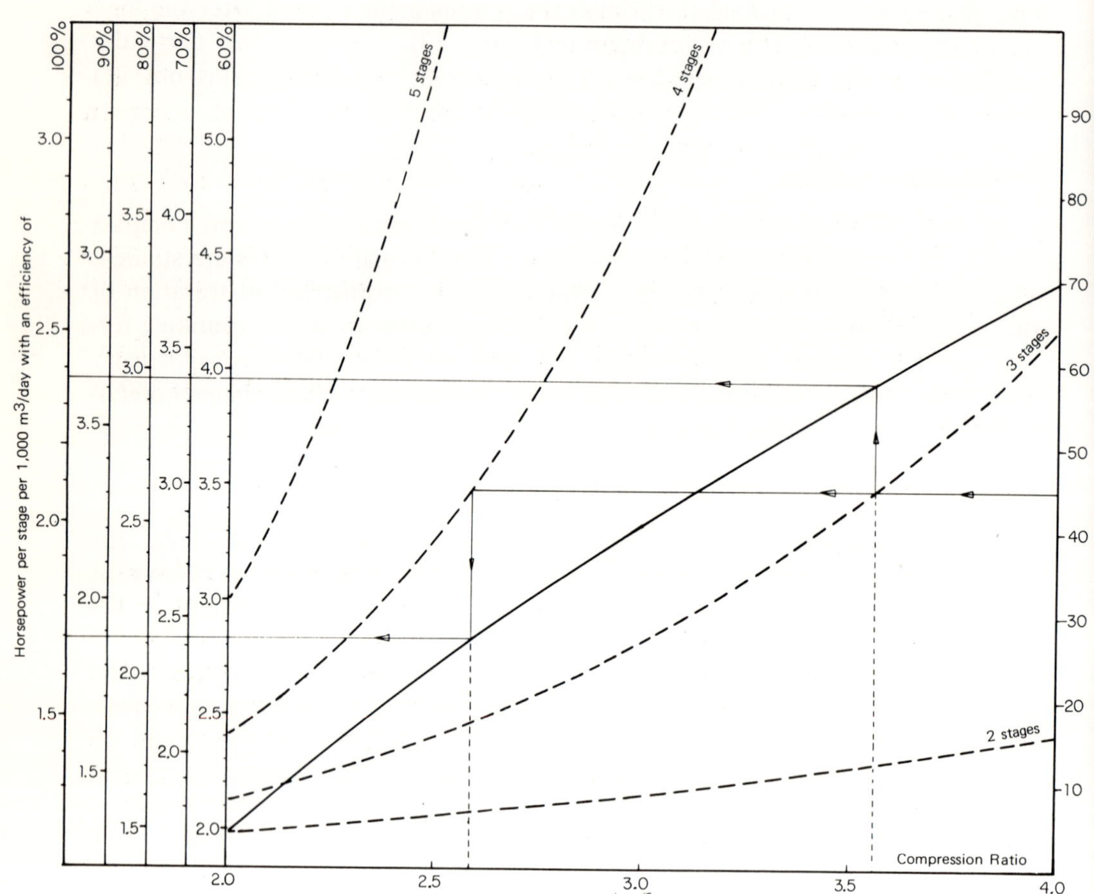

Fig. 63.31. Stage horsepower for air compressors.
(IFP study by B. Sahuquet).

(b) Injectivity: air must be injected into the well prior to ignition to induce a circulation path for the gas.

(c) Ignition: this often takes place by lowering a gas burner or electric ignitor in the well to the level of the formation. The energy required for ignition is generally arount 10^5-10^6 kcal/m^3 of formation.

If the oil in place is sufficiently oxidizable at bottom-hole conditions, spontaneous ignition may occur after several days of air injection. This phenomenon takes place more often in deep formations.

(d) Well completion: casing cement bonding must be good and the strings must be designed to cope with the possible thermal stresses.

(e) Monitoring and sampling: regular measurements of flowing temperature and produced gas composition are made, so that the progress of the combustion may be monitored.

(f) Separation of produced fluids: gas must be separated from the oil. Some oils form emulsions with water, and special treatment of these effluents is required.

2. Production mode: stimulation or continuous displacement

Stimulation

The combustion front is allowed to propagate only a few metres from the injection well, which results in the release of a considerable amount of heat in the formation. The same well is then put on production, the heat available increasing the productivity of the well. This technique may be of interest in deep formations where heat losses would be too great for efficient steam injection. In addition, in deep formations ignition often occurs spontaneously around the well bore after a period of air injection. This stimulation technique has been successfully attempted, but only on a very small number of wells. Successive cycles of limited combustion may be used, as in the huff and puff technique, but the length of the cycles is longer. The combustion phase usually lasts around 3 months, while the production phase may last more than 1 year.

Recovery by continuous displacement

The most typical case is that of an inverted five-spot pattern involving a production well at each corner of a square pattern and an injection well at its centre. The normal distance between wells is 100-400 m.

After ignition, the combustion front propagates from the injection well towards the production wells. To maintain the front, the amount of air injected is controlled so that there is always a sufficient supply of oxygen at the front. It is generally considered that this condition will be satisfied if the air flux at the front is such that its velocity u is equal to or greater than 4 cm/day (Ref. 26). If the surface area of the combustion front is s (m^2), the air flowrate

Q_a must be such that:

$$Q_a = aus \geqslant 4 \times 10^{-2}\ as \quad \text{sm}^3/\text{day} \qquad \text{(Eq. 63.31)}$$

where a is the air requirement (sm^3/m^3)

High flowrates are avoided because compression requirements become unreasonable and channeling may occur in the formation. An upper limit of 15 cm/day has been suggested for the velocity of the combustion front (Ref. 26). Thus we have:

$$15.10^{-2}\ as \geqslant Q_a \geqslant 4.10^{-2}\ as \quad \text{sm}^3/\text{day} \qquad \text{(Eq. 63.32)}$$

Since the surface area s of the front increases as the combustion front advances, the air flowrate is progressively increased so as to satisfy Eq. 63.32. The injection programme may be designed so that at all times Q_a remains as close as possible to its lower limit as defined by 63.31. However, some experts advise that injection should start with a period at which the frontal velocity is kept close to the upper limit of 15 cm/day, followed by a period in which the air flux gradually decreases to its lower limit (Ref. 26). The injection programme is in fact determined after consideration of both technical and practical criteria (in particular the most efficient use of the available compression capacity).

C. *Sweep efficiency and recovery factor*

There are two sweep efficiencies to be considered: that of the injected air and that of the high temperature combustion zone. The areal sweep efficiency of the air and the combustion front may reach 60% in homogeneous reservoirs. The vertical sweep efficiency of the air may be as high as 100% in thin formations, but is rather lower in thick formations because the air tends to segregate to the top of the reservoir. The vertical sweep efficiency of the combustion front is always lower than that of the air, because the combustion is extinguished near the top and bottom of the formation due to heat losses. The overall sweep efficiency of the combustion front hardly ever exceeds 30 to 40% of the volume defined by the production wells in the subject area.

However, it should be noted that the oil recovered comes both from the zones in which combustion takes place (from which all the oil not burnt is produced), and from the zones outside the path of the combustion front. The latter zones, heated by conduction, may have been swept by one of the displacing fluids (air, combustion gases, steam, etc.).

Thus under the most favourable conditions, in-situ combustion may yield recovery factors of 40-50%. This technique is therefore capable of providing large increases in recovery in heavy oil reservoirs, in which primary recovery would have only been of the order of a few percent.

D. Examples of field application of in-situ combustion

About a hundred in-situ combustion trials are referenced in the literature, and over twenty are beyond the experimental pilot stage. The characteristics of these tests are described in Table 63.32.

Several tests are currently in progress in the United States, principally in Texas, Louisiana and California. Trials have also taken place in Venezuela, Czechoslovakia, the USSR and Romania.

One Romanian test, started in 1964 in the Suplacu de Barcau field, is of particular interest (Ref. 28). Referring to the data given in Table 63.32 (No. 19), it can be seen that the conditions are most suitable for the application of in-situ combustion: highly viscous oil, shallow high permeability reservoir, medium thickness, high oil saturation and very low primary recovery forecast.

This dry combustion test started as a pilot scheme in a five-spot pattern 0.5 hectares in area. The successive positions of the combustion front are shown at monthly intervals on Fig. 63.32. Temperature surveys made in the observation wells were used to help determine the location of the front, which appears to have advanced regularly during the period shown. When the combustion front had reached all the pilot wells, air injection was continued and the front

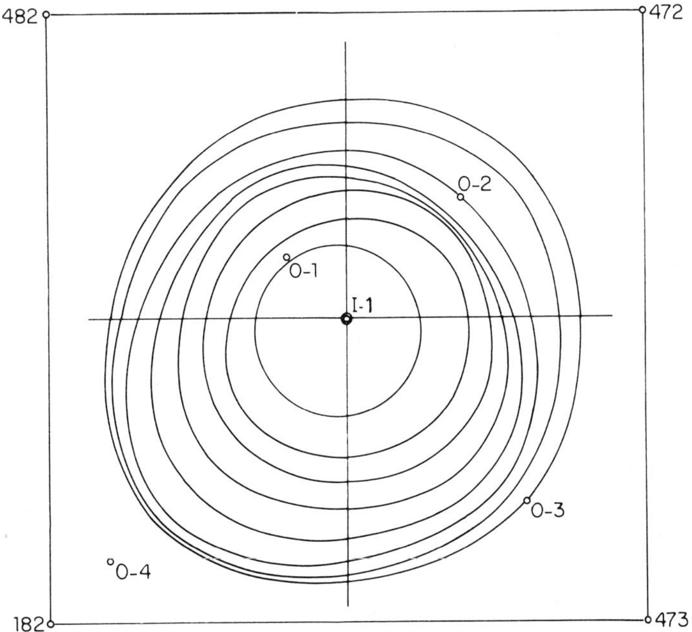

Fig. 63.32. Suplacu de Barcau field (Romania).
0.5 hectare pilot. Location of the combustion front shown at monthly intervals (Ref. 28).

TABLE 63.32

CHARACTERISTICS OF THE MAJOR FIELD TESTS OF IN-SITU COMBUSTION (Ref. 27)

No.	Location	Year	Formation	Pay (m)	Depth (m)	Porosity (%)	Permea-bility (mD)	Reservoir Temp. (°C)
1	S. Oklahoma	1952	Uncon. sand	6.1	55	29	2 300	16
2	Delaware-Childers, Okla	1949	Bartlesville sandstone	13	190-200	20.9	145	–
3	Parker Pool. Casey, III	1953	Pennsylvania sand	10.8	85	20.8	175	15.5
4	South Belridge, Calif.	1955	Tulare sand	9.1	215	37	8 000	30.5
5	S. Oklahoma	1953	Pontotoc sand	5.2	60	27.2	7 680	19
6	S.E. Kansas	1956	Bartlesville sand	2.7	250	20.3	85	25.5
7	Niitsu, Japan	1957	Sand	10	180-195	–	750	24
8	N. Tisdale, Wyo.	1958	Curtis sand	15.2	275	–	–	–
9	Shannon, Wyo.	1959	Shannon sand	10	290	23.3	250	20
10	E. Venezuela	1960	Uncon. sand	5.8	1 370	35	2 000-5 000	–
11	Midway-Sunset. Calif.	1960	Moco sand	39.4	640	36	1 575	52
12	Fry. III.	1961	Robinson sandstone	15.2	270-285	19.7	320	18
13	W. Newport, Calif.	–	B sand	76	245-640	37	1 000	41
14	Delaware-Childers, Okla.	1962	Bartlesville sand	13.9	185	20.6	118	–
15	N. Govt. Wells. Freer. Tex.	1962	Sandstone	6.1	710	32	1 800	49
16	Cox Penn. Okla.	1962	4th Deese sand	8-19	275-610	24	200-1 700	29.5
17	Bellevue, La.	1963	Nacatoch sand	22.5	100	–	–	–
18	Carlyle Pool, Kans.	1963	Bartlesville sand	10.7	260	25.3	2 050	23
19	Suplacu de Barcau Romania	1964	Sand	13.7	60-90	32	2 000	18
20	Pavlova Gora, URSS	1966	Sand	7	250	25	1 100	21
21	Delhifield, La.	1966	Tuscaloosa sand	2.5	1 040	31.2	1 069	57
22	E. Tia Juana, Venezuela	1967	Lagunillas sand	39	475	41	5 000	40
23	Trix Liz, Tex.	1968	Woodbine sand	6.1	1 120	30	–	–
24	Caddo Parish, La	1969	Nacatoch sand	4.9	315	35.2	606	27

* Oil content in litres per m^3 of porous medium.

TABLE 63.32 (Continued)

No.	Reservoir pressure (atm)	S_o (%)	S_w (%)	Oil		Test Pattern type	Area ($10^3 m^2$)	Primary recovery (%)	Average air injection rate ($10^3 sm^3$/day)	Average injection pressure (atm)
				Specific gravity	Viscosity cP (°C)					
1	4	60	40	0.943	7 413(16°)	3-5-spot	0.3	–	–	10
2	–	41	35	0.845	6	7-spot	21	25	23	–
3	–	42	32	0.897	50(15°)	5-spot	1.1	–	–	–
4	–	60	37	0.980	2 700(31°)	5-spot	10	10-15	100	16
5	5.5	64	35	0.944	5 000(19°)	7-spot	0.45	–	2.1	10
6	–	69	23	0.916	70(26°)	3×5-spot	240	5.3	10.2	65
7	–	25.4	74.6	0.946	178(25°)		0.15	40	4	11
8	–	(155)*	–	–	300	–	–	10	–	–
9	27	60	40	0.904	76(20°)	–	20	2	14	48
10	90	94	6	≈ 1	400	2-spot	–	2-7	28	93
11	70	75	25	0.969	110(52°)	–	–	4	170	51
12	1.4	68	20	0.884	40(18°)	5-spot	13	–	43	21
13	–	69	31	0.965	700(40°)	9×5-spot	156	10-15	14	13
14	–	23.6	30	0.860	6	5-spot	9	50 ?	54	49
15	5	45	41	0.922	10	5-spot	40	–	70	68
16	3.5-7	38-80	20-25	0.89-0.96	90-900(29°)	irregular	44	–	–	–
17	–	(193)*	–	0.940	–	9-spot	11	5	42.5	17
18	16	68	27	0.937	700(24°)	2×5-spot	17	2	6.8	17
19	4-10	78	22	0.959	2 000(18°)	5-9-spot	5-20-40	9	70	7
20	14.5	71	29	0.946	170(21°)	5-spot	15	9	18	34
21	–	30.6	30	0.825	3(57°)	5-spot	160	47	61	55
22	–	78	22	≈ 0.98	6 000	7-spot	46	–	–	–
23	–	–	–	0.922	–	4-spot	80	–	18	55
24	–	52	32	0.927	280(27°)	5-spot	11	–	15	20-40

advanced towards eight outer wells forming a larger square of 2 hectares centred on the same injection well I_1 (Fig. 63.33). The air/oil ratio was about 1 000 sm^3/m^3 during the test (Ref. 28).

Following the success of the extended pilot test, combustion was initiated at new injection wells: I_3, I_4, I_5, I_{12}, I_{13}, then at well 336. The location of the fronts at the end of 1970 is shown on Fig. 63.33. It can be seen that the fronts have advanced fairly evenly around the injection wells and that a satisfactory sweep efficiency has been obtained. The air/oil ratio was about the same as for the pilot scheme.

A recovery factor of more than 35 % has already been achieved in the zone produced by in-situ combustion (Ref. 29) and a recovery of about 50% is expected.

CONCLUSION

Thermal recovery methods rely on a number of complex phenomena of which some are not yet fully understood.

In practice thermal methods are very efficient, but they require heavy investment and more careful operating procedures than the standard methods of recovery. For these reasons, thermal methods currently have a rather restricted area of application. However, as the price of oil increases these methods will become economic for a greater number of fields in the future. Operating methods may also be developed to further improve the efficiency of thermal recovery (for example, by combining various recovery methods).

Thermal recovery is likely to play an important part in the exploitation of heavy oil reservoirs and tar sands, and possibly even oil shales. Up to now this type of reservoir has hardly been exploited, and the potential reserves involved are enormous.

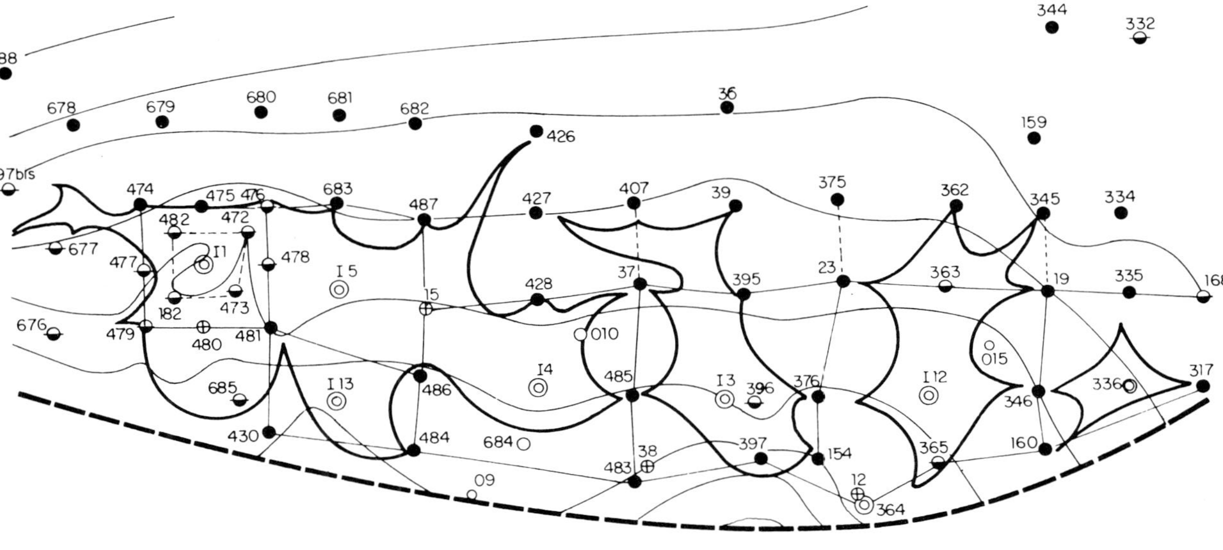

Fig. 63.33. Suplacu de Barcau field (Romania).
Combustion zone at the end of 1970.

REFERENCES

1 BIA P., BURGER J., COMBARNOUS M., SAHUQUET B. and SOURIEAU P., "Les méthodes thermiques de production des hydrocarbures". To be published by *Editions Technip* Paris.

2 FAROUQ ALI S.M., *Oil Recovery by Steam Injection.* Producers Publ. Cy., Bradford, 1970.

3 BRADEN W.B., "A Viscosity-Temperature Correlation at Atmospheric Pressure for Gas-Free Oils". *Annual Fall Meeting of Soc. Petroleum Engrs.,* SPE Paper n° 1 580, Oct. 1966.

4 POSTON S.M., YSRAEL S.C., HOSSAIN A.K.M.S., MONTGOMERY E.F. III and RAMEY H.J., Jr., "The effect of Temperature on Irreducible Water Saturation and Relative Permeability of Unconsolidated Sands". *Trans. Soc. Petroleum Engrs. AIME,* 249, 1970, p. 171-180.

5 WEINBRANDT R.M. and RAMEY H.J. Jr., "The Effect of Temperature on Relative Permeability of Consolidated Rocks". *Annual Fall Meeting of Soc. Petroleum Engrs.,* SPE. Paper n° 4 142, Oct. 1972.

6 LO H.Y. and MUNGAN N., "Effect of Temperature on Water-Oil Relative Permeabilities in Oil-Wet and Water-Wet Systems". *Annual Fall Meeting of Soc. Petroleum Engrs.,* SPE, Paper n° 4 505, Oct. 1973.

7 COMBARNOUS M. and PAVAN J., "Déplacement par l'eau chaude d'huiles en place dans un milieu poreux". *CR Troisième Colloque ARTFP,* Editions Technip, Paris, 1969, p. 737-757.

8 SOMERTON W.H., "Some Thermal Characteristics of Porous Rocks". *J. Petroleum Technol.,* 10-5, 1958, p. 61-64.

9 BURGER J. and SAHUQUET B., "Chemical Aspects of In-Situ Combustion. Heat of Combustion and Kinetics", *Soc. Petroleum Engrs. J.,*12, 1972, p. 410-422.

10 BONNIER J.M. and de GAUDEMARIS G., "Stabilité thermique des hydrocarbures". *Rev. Inst. Franç. du Pétrole,* 17, 1962, p. 852-882.

11 HARMSEN G.J., "Oil Recovery by Hot-Water and Steam Injection". *Proc. 8th World Petroleum Congr.,* 3, 1971, p. 243-251.

12 SOURIEAU P., "Comparaison de la stabilité des déplacements par fluide chaud (eau ou vapeur d'eau) et par l'eau froide". Not IFP D.43.1, n° 20, 1973.

13 MARX J.W. and LANGENHEIM R.H., "Reservoir Heating by Hot Fluid Injection". *Trans. Soc. Petroleum Engrs. AIME,* 216, 1959, p. 312-315.

14 RAMEY H.J. Jr., "Wellbore Heat Transmission". *J. Petroleum Technol.,* 4, 1962, p. 427-435.

15 HALL A.L. and BOWMAN R.W., "Operation and Performance of the Slocum Field Thermal Recovery Project". *J. Petroleum Technol.,* 25, 1973, p. 402-408.

16 ARMSTRONG T.A., "Thermal Recovery, where Steam pays off ?". *Oil and Gas Journal,* 64-14, 1966, p. 127-135.

17 DIETZ D.N., "Hot Water Drive". *Proc. 7th World Petroleum Congr.,* 3, 1967, p. 451-457.

18 AFOEJU B.I., "Conversion of Steam Injection to Waterflood East Coalinga Field". *J. Petroleum Technol.*, 26, 1974, p. 1227-1232.

19 ALEXANDER J.D., MARTIN W.L. and DEW J.N., "Factors Affecting Fuel Availability and Composition during In-Situ Combustion". *J. Petroleum Technol.*, 14, 1962, p. 1154-1164.

20 WEIJDEMA J., "Zur Oxydationskinetik flüssiger Kohlenwasserstoffe in porösen Medien in Bezug auf unterirdische Verbrennung". *Erdöl u. Kohle*, 21, 1968, p. 520-526.

21 BURGER J. and SAHUQUET B., "Combustion à contre-courant. Interprétation d'essais par modèle numérique unidirectionnel". *Rev. Inst. Franç. du Pétrole*, XXVI, 1971, p. 399-422.

22 WILSON L.A., REED R.L., REED D.W., CLAY R.R. and HARRISON N.H., "Some effects of Pressure on Forward and Reserve Combustion". *Soc. Petroleum Engrs. J.*, 3, 1963, p. 127-137.

23 DIETZ J.N., "Wet Underground Combustion. State of the Art". *J. Petroleum Technol.*, 22, 1970, p. 605-617.

24 BURGER J. and SAHUQUET B., "Laboratory Research on Wet Combustion". *Annual Fall Meeting of Soc. Petroleum Engrs.*, SPE Paper n° 4 144, Oct. 1972, *J. Petroleum Technol.*, 25, 1973, p. 1137-1146.

25 BUXTON T.S. and CRAIG F.F. Jr., "Effect of Injected Water-Air Ratio and Reservoir Oil Saturation on the Performance of a Combination of Forward Combustion and Waterflooding". *Amer Inst. Chem. Engrs., 71st Nation. Meeting*, Feb. 1972. *Amer. Inst. Chem. Engrs., Symp. Series*, 69, 127, 1973, p. 27-30.

26 NELSON R.W. and Mc NIEL J.S., "In-Situ Combustion Project". *Oil and Gas Journal*, 59-23, 5 June 1961, p. 58-65.

27 FAROUQ ALI S.M.., "A Current Appraisal of In-Situ Combustion Field Tests". *J. Petroleum Technol.*, 24, 1972, p. 277-486.

28 ALDEA G., PETCOVICI V., DUMITRESCU H. and PALADA T, "Aplicarea experimentala a metodei de exploatare prin combustie subterana in Republica Socialista Romania". *Petrol. si Gaz*, 19, 1968, p. 26-36.
PETCOVICI V., "Considérations sur les possibilités du contrôle du front de combustion dans un processus de combustion in situ sur champ". *Rev. Inst. Franç. du Pétrole*, XXV, 1970, p. 1 355-1 374.

29 BURGER J., ALDEA G., CARCOANA A., PETCOVICI V., SAHUQUET B., and DELYE H., "Recherches de base sur la combustion in situ et résultats récents sur champ". *Proc. 9th World Petroleum Congr.*, 3, 1975, p. 279-289.

7

other methods of enhanced recovery

by C. BARDON and M. LATIL

71. INTRODUCTION

We have seen that water or gas injection in an oil reservoir results in a rather less than perfect recovery. At the end of the exploitation a substantial amount of oil remains in place due to:

(a) A partial sweep of the reservoir.
(b) Oil trapped by capillary forces in the invaded zones.

Research into more efficient enhanced recovery methods therefore has one of the following objectives:

(a) To improve the sweep efficiency, by reducing the mobility ratio between injected and in-place fluids.
(b) To eliminate or reduce the capillary forces and thus improve displacement efficiency.
(c) To act on both phenomena simultaneously.

The various methods suggested or tested for the improvement of enhanced recovery may be grouped under these three headings as shown in Table 7.11. Thermal methods and miscible displacements, which have already been covered in this volume, are shown together with methods which are often still at the research or pilot stage and which will be briefly covered in this chapter.

72. THE USE OF POLYMERS

The polymers in question are those water-soluble polymers which are insoluble in oil or alcohol. They have molecular weights in the millions and are used in aqueous solutions at concentrations of 0.1 to 1 ppm.

Three principal types of polymer are used: polyacrylamides, polysaccharides and ethylene polyoxide. The first two would appear to have been the only polymers tested in the field with the polyacrylamides being the most popular.

Polymer solutions have the advantage of being very viscous even when highly diluted. Currently, viscosities of the order of 10 to 100 cP are obtained.

The mobility of solutions in porous media often proves to be lower than would have been predicted on the basis of the increased viscosity alone. Thus for polymer solutions the idea of a "mobility reduction"[1] has been introduced. This is defined as the ratio between the mobility of water and the mobility of water containing polymers.

$$R \text{ or } F_r = \frac{M_w}{M_p}$$

TABLE 7.11

OTHER METHODS OF ENHANCED RECOVERY

Basic Principle	Methods Used
Improvement of sweep efficiency	Polymer solutions Water-gas foam injection
Improvement of displacement efficiency	Miscible fluids (alcohol, LPG or rich gas, dry gas) Surfactants Wettability reversing agents
Improvement of both sweep efficiency and displacement efficiency	Micro-emulsions (e.g. "Maraflood") Alternating water-gas injection Hot water Steam Carbon dioxide Bacterial action Forward combustion

The value of R can only be determined by laboratory experiment and the improvement in recovery by polymer injection cannot be calculated from rheological data alone. Experimental study of the displacement of reservoir oil by the polymer solution, using core samples of the porous medium, enable both the mobility reduction R and the microscopic displacement efficiency E_d to be de-

(1) Also known as the "resistance factor" or the "apparent viscosity in a porous medium".

termined. Knowing the mobility of the polymer solution in the porous medium, the standard curves of Caudle and Witte (Figs. 15.21 and 15.22) can be used to estimate E_s. Thus all the elements of the calculation are available. However, this method of calculation is subject to error due to the dilution of the solution and the retention of polymer in the porous medium. To go further a numerical model is required.

The passage of a polymer solution through a porous medium causes a permanent reduction in the mobility to water. The mobility to oil however remains practically constant.

The rheological behaviour of polymer solutions is complex. These non-Newtonian solutions behave in a pseudo-plastic manner in the free state, their viscosity decreasing with increasing constraint. However, when dilated in porous media their viscosity increases as the speed of circulation increases (*cf* Appendix 7.2). Various hypotheses concerning the structure of polymer solutions have been proposed to explain these apparently contradictory properties. In particular, it has been noted that polymer solutions may not be considered to be continuous solutions.

It seems well established that the viscosity of polyacrylamide solutions is greatly reduced in the presence of salts. Thus the possibility of the polymer solution becoming contaminated by interstitial water may be a problem. In contrast, the viscosity of ethylene polyoxide solutions does not depend on salinity. Very little data is available on polysaccharide solutions.

Polymer solutions become degraded with time, the ageing being principally due to the presence of oxygen. Thus in the practical application of polymer solutions either sulphites or formaldehyde are added, the latter having an additional bactericidal action([1]). Thermal degradation of the commonly used commercial products starts at about 125° C and is total at 175° C. In the study of polymer applications stability tests should be made at reservoir conditions.

Polymer molecules form long chains which may be broken if violently agitated. The service companies have developed appropriate equipment and methods.

In porous media there is interaction between the solid matrix and the macromolecular chains, causing a loss of polymer both by adsorption and by physical trapping within the pores. The amount of polymer retained by the matrix is a function of the natures of the porous medium and the polymer, and the concentration of the polymer solution. Typically the retention is of the order of 10 to 200 μg of polymer per g of rock. During displacement of oil by polymer solutions in a porous medium containing a residual water saturation, a water bank

([1]) An alternative bactericide used is mercuric chloride at 10 ppm.

forms between the oil and the polymer solution. This water bank contains both connate water and injected water whose polymer content has been lost to the matrix.

The effect on oil recovery is negligible in the case of a homogeneous medium and unidimensional flow, but it becomes significant in a heterogeneous medium or a two dimensional homogeneous medium. There is a marked increase in oil recovery both at breakthrough and at abandonment.

The commercial interest in the process is evidenced by the number of field tests, both pilots and full-scale, in which the producting water-cut has been reduced. Over 50 tests have been performed so far, and most have been successful.

This method is recommended for reservoirs in which sweep efficiency is of paramount importance, in particular those heterogeneous reservoirs containing relatively viscous oil (20 to 200 cP) and whose temperature does not exceed 120° C.

The profitability of polymer flooding is uncertain, the major cost being that of the polymer itself, used in a slug of around 10 to 30 % of the reservoir pore volume. In 1969 the following costs applied:

(a) Polymer price: $1.25 per pound.
(b) Cost of secondary oil recovered: $0.50 per bbl.

By comparison, the price of crude oil was about $2/bbl in 1971 and $12/bbl in 1974.

73. FOAM INJECTION

Foams are accumulations of gas bubbles, separated from each other by thin films of liquid. They have the property of having a higher viscosity than that of the gas or liquid of which they are composed (see the work of Fried).

The injection of foam into a reservoir thus takes place at a lower mobility ratio than that of gas or liquid injection alone. The process was first proposed in the late fifties (see Bond and Helbrook's patent of 1958).

The "quality" of the foam is defined as the ratio of the contained gas volume to the total foam volume. Raza and Marsden have shown that in practice the maximum foam quality which can be achieved is 0.96, since a minimum of 4% liquids is required to produce the thin liquid films.

Foams with a quality higher than 0.8 are termed "dry" while those with a quality lower than 0.7 are termed "wet".

Wet foams are characterized by the presence of large cylindrical gas bubbles, separated by slugs of iiquid.

Dry foams exhibit better dispersion of the two phases and are thus more stable. Most laboratory research has been performed using high quality dry foams.

The injection of foam into a porous medium creates a large number of resilient interfaces, which exert a piston-like force on the oil to be displaced (Jamin effect). The process is highly efficient since the foam first finds its way into the largest pores, in which it tends to obstruct further flow. The smaller pores are thus invaded next, and so on until the entire permeable section has accepted the foam. The vertical sweep efficiency is thus much improved.

There are no theoretical or empirical methods by which the optimum size of a foam slug may be determined, whether for linear or more complex systems. Field pilot tests must be run in order to obtain a better idea of the required size of foam slug.

Laboratory experiments have shown that using selected surfactants foams may remain stable for more than a month at rest, and even longer when flowing in a low permeability porous medium ($k < 1$ mD).

Since foam stability is not perfect, most practical methods require continuous foam injection.

Many studies have shown that oil behaves as a partial breaking agent for many types of foam. The exact behaviour of a foam cannot be accurately predicted if laboratory tests on core samples at reservoir conditions have not been performed. Even then, the scaling factors needed to extrapolate from laboratory to field conditions are not well defined.

Various relationships giving displacement efficiency as a function of injected pore volume and mobility ratio were shown in Chapter 1. In all cases it was assumed that the mobility of the displacing phase behind the front remained constant. When considering the case of a slug of foam driven by gas these conditions no longer apply, but the sweep efficiency derived from the mobility ratio between the dry gas and the oil will be a conservative estimate. The larger the slug of foam the larger the error in estimates made in this way.

To illustrate the severe reduction in permeability to gas which may occur due to the use of foam we should consider the results of a study by Bernard and Holm. Using a core sample with a absolute permeability of 3 890 mD, they found that the permeability to gas was practically zero at a foaming agent concentration of only 1%.

This practically complete arrest of the gas flow may become a problem. Although it is clearly desirable to reduce the permeability to gas, it is only necessary to achieve a favourable gas-oil mobility ratio, i.e. around 1.

Foaming agents must therefore be used at very low concentrations.

If field application of foam drive is preceded by laboratory tests, and if it is most carefully controlled, economically acceptable rates of oil recovery should result.

74. THE USE OF SURFACTANT SOLUTIONS

Although first proposed in the 1920's, attempts to use surfactants in aqueous solutions have so far met with failure.

However, interest in the technique has been recently revived by several American petroleum companies, as witnessed by papers presented at the April 1972 SPE meeting in Tulsa. Most research workers consider that the primary cause of the failures noted to date was that the interfacial tension was not reduced sufficiently to have an effect on the trapped oil.

It would appear that it is necessary to reduce the interfacial tension between the oil and the slug of surfactant-bearing water to the order of 0.01 to 0.001 dyn/cm and to be able to maintain this value for the duration of the displacement.

Laboratory studies have shown that the required large reduction in interfacial tension can only be achieved in a very narrow range of sodium chloride concentration (0.2 to 0.3 mole/litre). Thus it is necessary to inject water of suitable salinity before injecting the surfactant solution. This pre-flush displaces the formation brine so that it does not come into contact with the surfactant solution.

The surfactants used to date are petroleum sulphonates derived from crude oil. They are inexpensive, easily obtainable in large quantities and have high interfacial activity.

Much effort has been made to understand and control the action of sulphonates. It has been found that there is correlation between displacement efficiency and the equivalent weight of a sulphonate. (The equivalent weight is the ratio of the molecular weight to the number of sulphonate groups present in the molecule).

Sulphonates with high equivalent weights cause the greatest reduction in interfacial tension but are unfortunately insoluble in water and are readily adsorbed. The first problem is resolved by mixing these sulphonates with those of lower equivalent weight, thus achieving a compromise between solubility and reduction of interfacial tension. The second problem is attacked by adding a slug of mineral compounds which attach themselves to the adsorption sites in the porous media in preference to the surfactant.

The optimum average molecular weight of sulphonate mixtures is between 400 and 450 with the equivalent weight between 375 and 475.

Generally, in order to ensure that the mobilities may be well controlled, the surfactant slug is driven by a polymer slug.

In addition, various mineral additives are used with the surfactant slug to protect it against mineral salts in the formation water by the precipitation or sequestration of the divalent cations. The most popular additives are ammonia, sodium carbonate and sodium triphosphate.

75. MICRO-EMULSIONS. THE MARAFLOOD PROJECT

This technique was developed by the Marathon Oil Company and consists of the use of water-oil emulsions of which the discontinuous phase is in the form of extremely small droplets (10^{-4} to 10^{-6} cm). The microscopic size of the droplets is achieved by the use of a large percentage of a suitable surfactant and a certain amount of alcohol. Micro-emulsions are also known as "micellar solutions".

An example of the composition of a micro-emulsion is as follows:

	%
surfactant	10-15
alcohol	1-3
water	15-45
oil	40-70

The major interest in these fluids is that it is possible to pass continuously from the case of a micro-emulsion in which oil is the continuous phase to one in which water is the continuous phase. This then suggests that a true miscible displacement of oil by water may be possible.

The principal properties of micro-emulsions are:

(a) Reasonable compatibility with the various reservoir fluids. A micro-emulsion in which oil is the continuous phase is compatible with crude oils and this can be regarded as miscibility in the normal sense of the word. Contact with water creates a continuous zone in which the micro-emulsion tends progressively to have a continuous water phase. The zone in which this emulsion reversal takes place has an extremely high viscosity.

The presence of salts may completely change the properties of micro-emulsions.

(b) Complex rheological properties: e.g. non-Newtonian behaviour, the apparent viscosity depending on the many parameters which define the micro-emulsion.

Field application is extremely complex, the slug of micro-emulsion with a continuous oil phase being followed by a viscous slug and then by water injection. The viscous slug, generally a polymer solution, helps to avoid contamination of the very expensive micro-emulsion by the following water.

On average, mobility control is achieved with the following injected volumes:

(a) Micro-emulsion slug: 5 to 15% pore volume.
(b) Polymer slug: 30 to 60% pore volume.

The economics of this technique, still at the laboratory stage despite industry interest, have not yet been firmly established. At the least expensive, the cost price of micro-emulsions is around $ 6/bbl, which would make any supplementary production obtained far too costly.

American companies are currently keeping a close eye on Marathon's work, especially the field trials now in progress.

76. THE USE OF CARBON DIOXIDE

Carbon dioxide may be used as a gas, dissolved in water or in an alternating slug scheme (Refs. 15 to 18).

The very high solubility of carbon dioxide in oil and to a lesser extent in water results in:

(a) A large reduction in oil viscosity and a small increase in water viscosity. This results in a significant improvement of oil mobility in the reservoir.

(b) Swelling of the oil by some 10 to 20%, depending on its type and saturation pressure.

(c) A reduction in oil density. This lessens the effect of gravity segregation during the injection of gaseous carbon dioxide.

(d) A lowering of the interfacial tension. With CO_2 in the gaseous state at high enough pressure, miscibility with oil may be achieved.

(e) Chemical action on carbonate or shaly rocks.

76.1. The effects of CO_2 on fluids

A. Oil

Figure 76.11 shows that oil viscosity reduction depends to a large extent on its saturation pressure. For any given saturation pressure the viscosity reduction is relatively greater the higher the original oil viscosity.

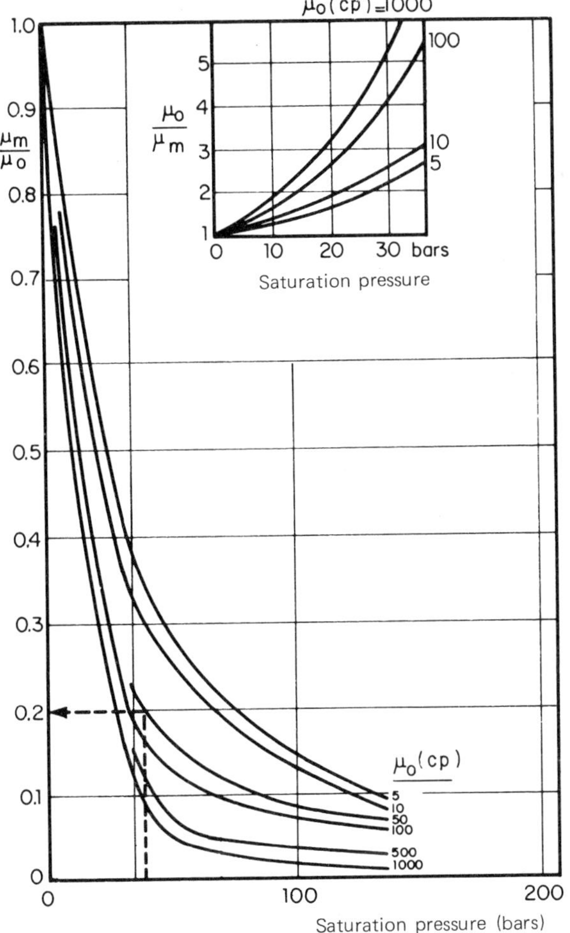

Fig. 76.11. Viscosity of oil saturated in CO_2 at 49°C (From Simon and Graue, Ref. 19).

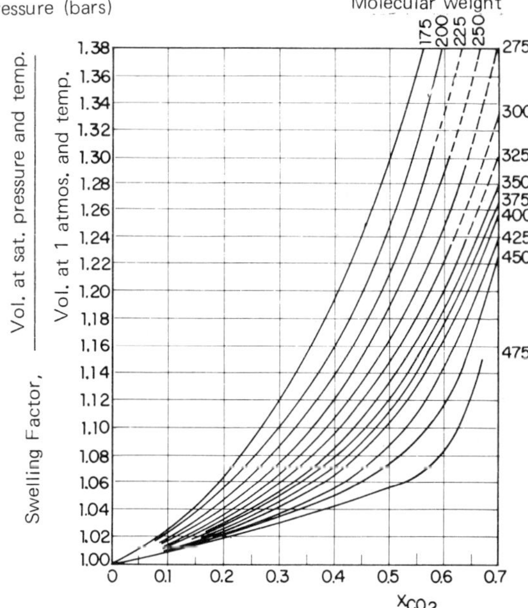

Fig. 76.12. Crude oil swelling coefficient as a function of the CO_2 content (mole percent) and the molecular weight of the oil (From Simon and Graue, Ref. 19).

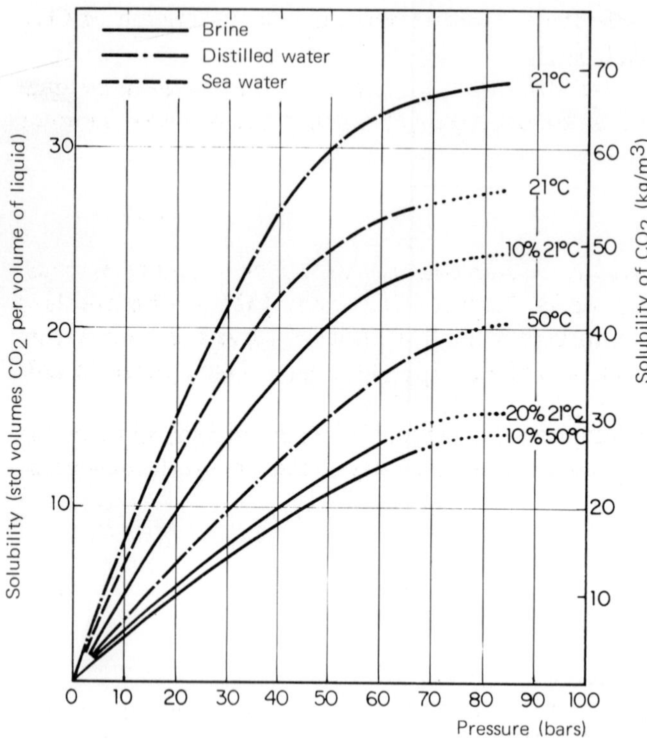

Fig. 76.13. Solubility of CO_2 in fresh and saline waters.

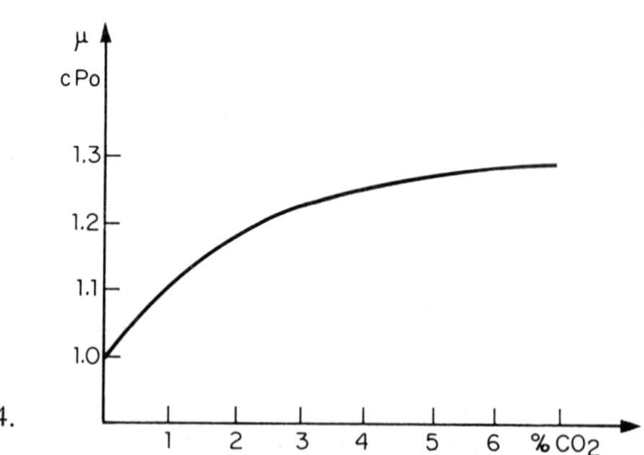

Fig. 76.14.

Nota. Simon and Graue (Ref. 19) have produced additional figures which may be used in conjunction with Fig. 76.11 to determine μ_m / μ_0 at temperatures other than 49° C.

Figure 76.12 gives the crude oil swelling coefficient (volume of CO_2 saturated crude/volume of gas-free crude at the same temperature) as a function of CO_2 content and oil properties (Ref. 18).

Oil swelling helps to improve the recovery factor, since for a given residual oil saturation, the mass of oil abandoned is lower than if the oil was more or less gas-free.

B. Water

CO_2 is also soluble in water, but in contrast to oil its solubility is much less sensitive to pressure. Above 70 bar very little extra CO_2 can be dissolved in water. The solubility is temperature sensitive, decreasing as the temperature increases. Salinity also has an effect on solubility, CO_2 being more soluble in fresh water than in brine (Ref. 20) (Fig. 76.13).

The increase in water viscosity obtained (Ref. 21) (Fig. 76.14) improves the water-oil mobility ratio and is an obvious advantage in flooding with carbonated water.

76.2. The effects of CO_2 on rock (Ref. 22)

The principal effect of CO_2 is that due to the action of carbonic acid, formed in solution with water according to the following equilibrium relationship:

$$CO_2 + H_2O \rightleftarrows H_2CO_3$$

Its effect on shaly rocks is a stabilising one due to the reduction in pH. This stops the shales from swelling and causing blockage of the porous medium.

Its effect on calcareous rocks is to improve injectivity by partially dissolving the rock, according to the following reactions:

$$H_2CO_3 + CaCO_3 \rightleftarrows Ca(HCO_3)_2$$

$$H_2CO_3 + MgCO_3 \rightleftarrows Mg(HCO_3)_2$$

The bicarbonates thus formed are quite soluble in water. Figures 76.21 and 76.22 show the solubility of calcium and magnesium carbonates in carbonic acid. This effect may give rise to an increase in the permeability of the carbonate rock, especially around the wellbore. Figure 76.23 shows the results of flooding a core sample of a calcareous rock with carbonated water. After the injection of 40 pore volumes its permeability had increased three-fold (Ref. 23).

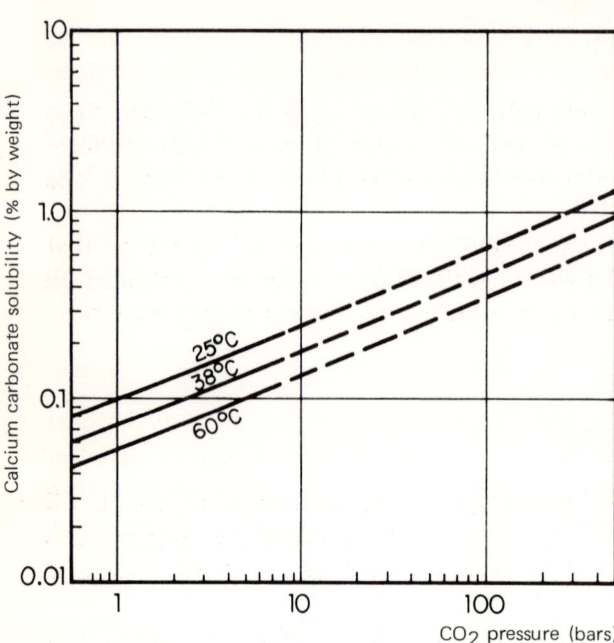

Fig. 76.21. Solubility of calcium carbonate in carbonic acid. (Ref. 22).

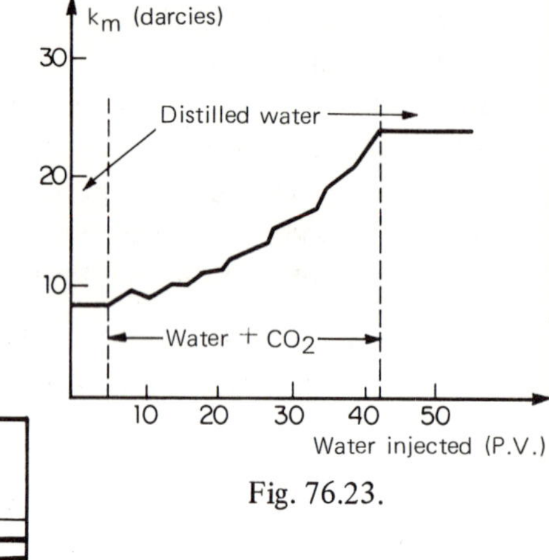

Fig. 76.23.

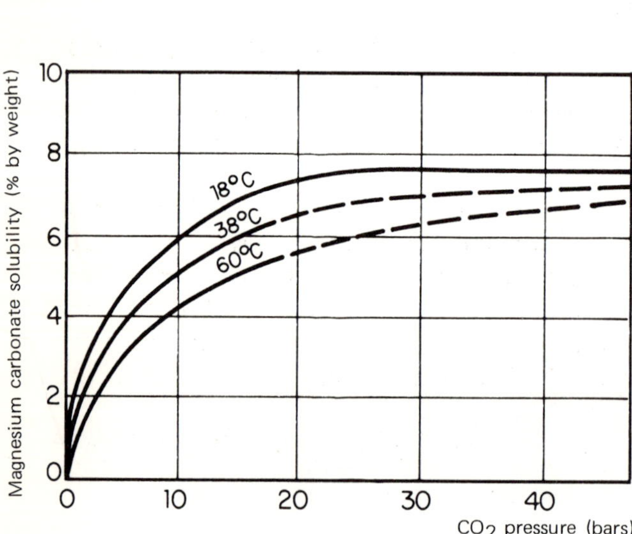

Fig. 76.22. Solubility of magnesium carbonate in carbonic acid (Ref. 22).

76.3. Non-miscible displacement by CO_2

CO_2 injection affects relative permeabilities by changing the fluid viscosities and interfacial tensions. Figures 76.31 and 76.32 show typical relative permeability curves for gas-oil and carbonated water-oil systems, as obtained in the laboratory (Ref. 24).

The residual oil saturation obtained by CO_2 injection is lower than that obtained by using natural gas. This is in addition to the already mentioned oil swelling that occurs, and provides an even greater improvement in the recovery factor.

76.4. Miscible displacement by CO_2

The miscibility achieved may be partial or complete depending on the oil characteristics. The solubility of CO_2 in oil is a function of the oil properties and the pressure and temperature, as shown on Figs. 76.41 and 76.42. In general the solubility of CO_2 is greatest in light oils (Ref. 15).

As fig. 76.42 shows, with some crudes a limiting solubility is reached at a certain pressure (Ref. 22).

In the case of some light oils thermodynamic miscibility may be achieved at pressures of the order of 140 to 210 bar (2 000-3 000 psig). This is around 50 to 100 bar less than in the case of high pressure gas injection.

With very viscous oils the miscibility pressure can never be reached. However, the CO_2 dissolved in the oil has a direct effect on the properties of the mixture, and the viscosity reduction thus obtained is obviously beneficial.

- Formation of the miscible bank

The injected carbon dioxide mixes with the oil by dissolution, the process being quite complex. During displacement of the CO_2 within the porous medium there is a large contact area between gas and oil. A rapid mass transfer between the oil and the CO_2 takes place by fractionation of the oil (Fig. 76.43).

The frontal part of the mixing zone thus becomes progressively richer in light hydrocarbon fractions. If the oil contains a significant quantity of methane it may be extracted from the oil and travel just ahead of the CO_2 front. This phenomenon has been confirmed by laboratory experiments on full diameter core samples. The formation of a methane bank between the oil and the CO_2 saturated zone was observed when the injection pressure was lower than the miscibility pressure of methane (Ref. 25).

In the mixing zone, the intermediates and CO_2 make the oil significantly lighter.

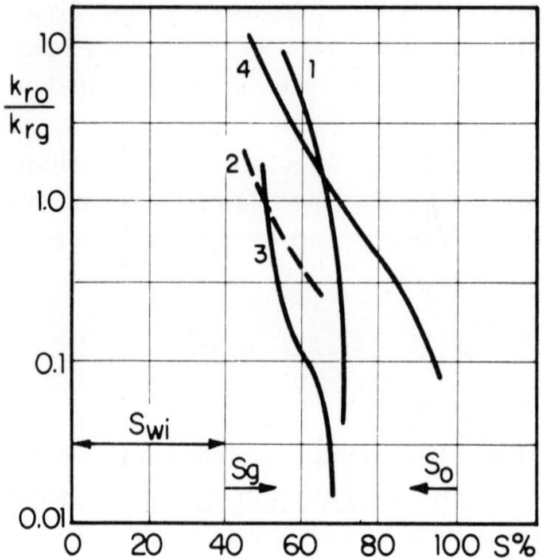

S_{wi} = 40 % S_o = 60 %
Ø = 38-41 % k_w = 18-28 mD
k_{gc} 46-72 mD

1. Displacement with natural gas
   p = 130 atm T = 82° C
2. Displacement with CO_2
   p = 130 atm T = 82°C
3. Displacement with CO_2
   p = 20 atm T = 82°C
4. Displacement with CO_2
   p = 130 atm T = 40°C

Fig. 76.31. Gas-oil relative permeability curves (Ref. 24).

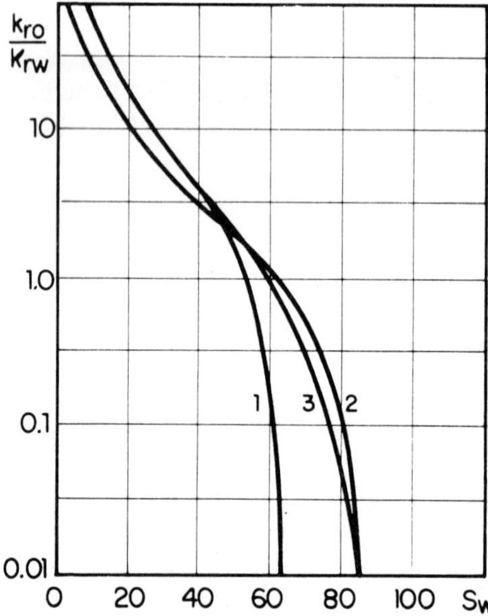

S_{wi} = 0 S_o = 100 %
Ø = 36.5 % K_{ge} = 37-47 mD
T = 82 C° K_o = 30-35 mD
p = 130 atm

1. Displacement of oil by water
2. Displacement of oil containing CO_2 by water containing CO_2 (R_{sw} = 20.6)
3. Displacement of oil without gas by water containing CO_2 (R_{sw} = 20.6)

Fig. 76.32. Water-oil relative permeability curves showing the effect of dissolved CO_2 (Ref. 24).

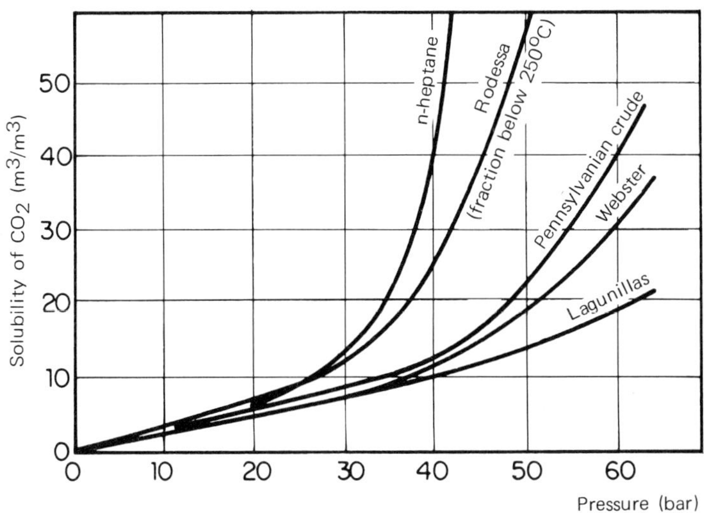

Fig. 76.41. Solubility of CO_2 in various crudes and crude fractions at c.24 °C) (Ref. 16).

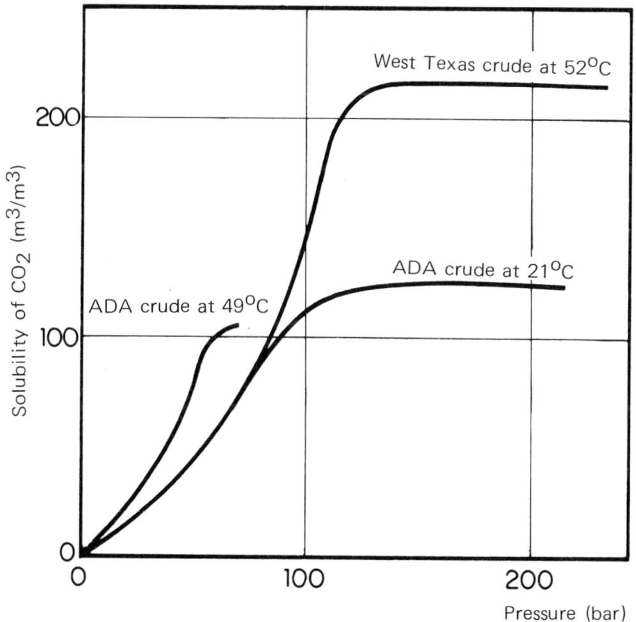

Fig. 76.42. Solubility of CO_2 in various crudes (Ref. 22)

Finally, behind the front the oil becomes progressively heavier, and although saturated in CO_2 it has a relatively low mobility due to the loss of all its light ends.

- Ternary diagram

There are two ways of constructing the ternary diagram depending on whether the CO_2 is associated with the methane or with the intermediates.

Using the first method (Fig. 76.44) the miscibility conditions applying to the injection of CO_2 or methane can be compared (Ref. 25).

With the second method (Fig. 76.45) the CO_2 point on the ternary diagram is located according to its molecular weight along the line joining the C_1 and C_2-C_6 points. It can thus be compared to ethane or propane. Its behaviour is in this method analogous to that of an enriched gas.

In reality neither of the two methods is entirely satisfactory, and to depict the effect of CO_2 more accurately a tetrahedral representation is required, using the following groupings: $C_1 - N_2$, CO_2, $C_2 - C_6$, C_{7+} (Fig. 76.46).

76.5. Field applications

The practical success of CO_2 injection depends on:

(a) The oil characteristics.

(b) The part of the reservoir effectively contacted.

(c) The attainable pressure.

(d) The availability and cost of the CO_2.

The relatively high cost of CO_2 is an important factor, and is the reason why the commercial development of the method is not as far advanced as might have been expected. Some field tests have been carried out however, in particular in Hungary (Ref. 26) where large carbon dioxide reservoirs have been discovered, often above or below oil reservoirs and at high enough pressure as to require little or no compression. In the United States the largest project yet undertaken is the SACROC project in Texas (Ref. 27) which started in 1972. The realisation of this project over the entire Kelly-Snyder field required heavy capital investment, in particular for the construction of a 100 km CO_2 pipeline. The test in the Mead-Strawn field should also be noted, where a definite profit was obtained (Ref. 28).

Injected gas	Injected gas + heavy fractions of residual oil	gas enriched by evaporation of the oil	oil enriched by intermediates	virgin oil

	CO_2	CO_2 + gaseous hydrocarbons + oil in equilibrium with the gas	enriched oil	
	heavy residual oil			
		irreducible water		

Fig. 76.43.

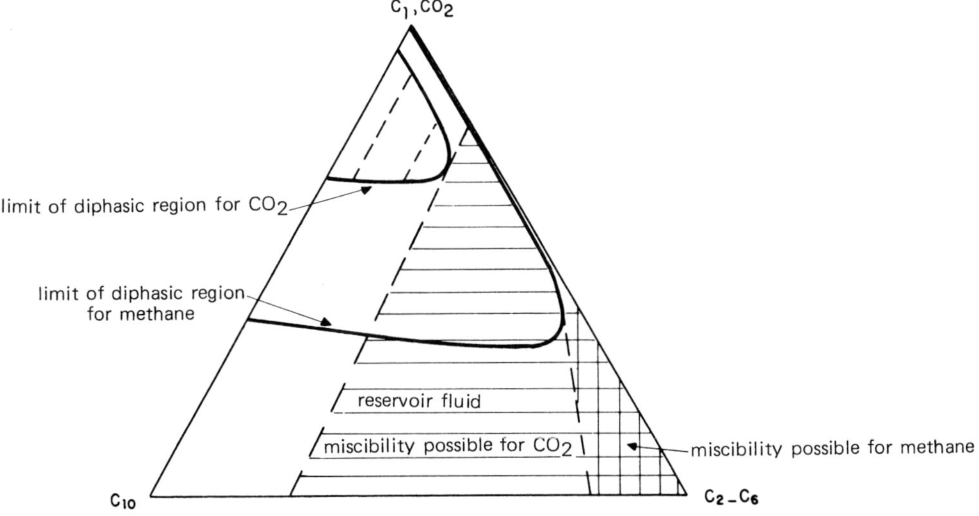

Fig. 76.44. The behaviour of methane and CO_2 during high pressure injection.

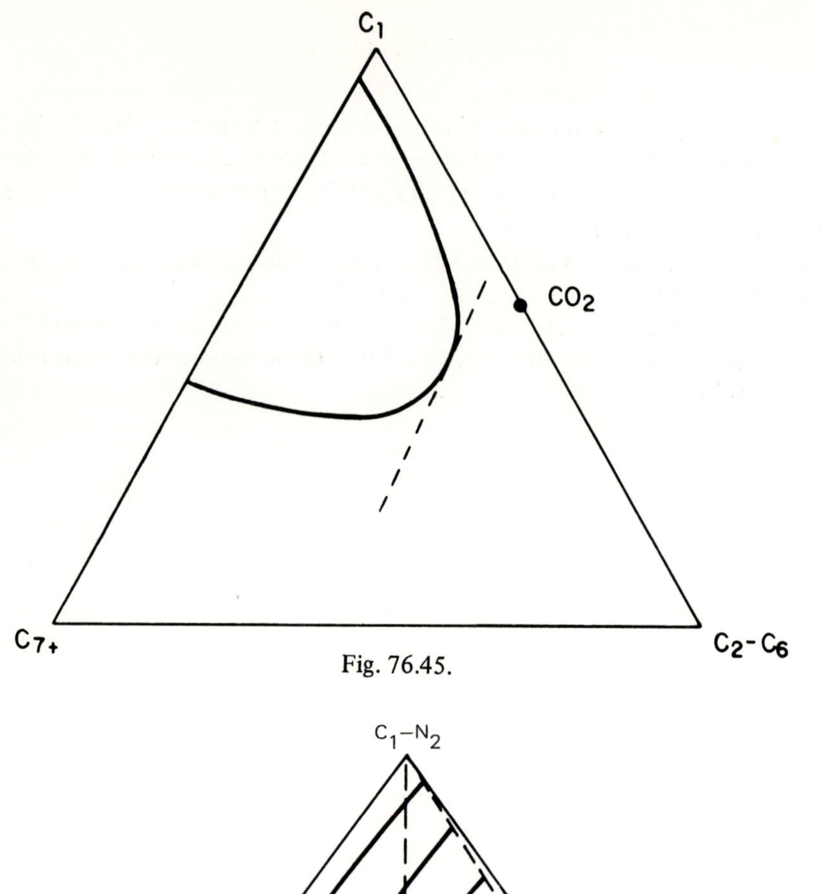

Fig. 76.45.

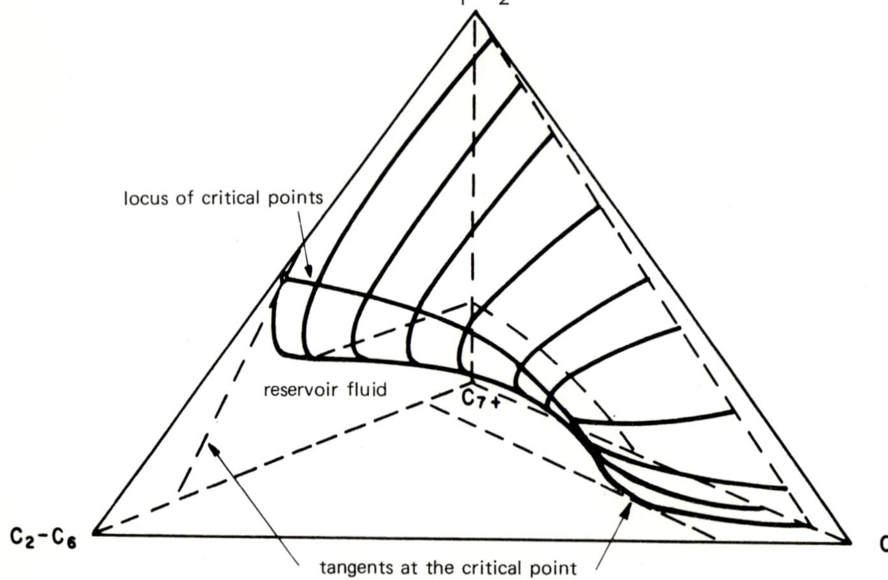

Fig. 76.46. The behaviour of a four-component system including CO_2.

77. CONCLUSIONS

At present, a wide variety of enhanced recovery methods exist of which none is universal. Moreover, no single method could ever be universal because of the vast range of reservoir and economic conditions encountered and the complex nature of the problems involved.

Finally, the capital investments required are high and are only recovered towards the end of the life of the reservoir.

It is evident that serious research into enhanced recovery methods must be continued both for the development of existing methods and for the promotion of new techniques.

APPENDIX 7.1

AN ANALYSIS OF FIELD TESTS OF ENHANCED RECOVERY METHODS OTHER THAN WATER OR GAS FLOODING IN THE USA IN 1971

Method	Analysis				
	Number of tests	%	Pilot tests	Full scale tests	%
Combustion	46	31.0	25	21	24.4
Steam soak	31	21.0	2	29	33.8
Steam flood	24	16.2	16	8	9.3
Polymer flood	16	10.8	8	8	9.3
Micro-emulsions	5	3.3	5	0	0
Miscible displacement by:					
Hydrocarbons	22		4	18	
Flue gas	3	17.7	1	2	23.2
CO_2	1		1	0	
Total	148	100.0	62	86	100.0

(*Source* : *Oil and Gas Journal,* May 3, 1971)

APPENDIX 7.2

The rheology of polymer solutions

The rheological properties of polymer solutions are extremely complex. In this Appendix only the fundamental viscoelastic and laminar flow properties will be discussed.

1. Viscoelasticity and relaxation time

Aqueous polymer solutions are viscoelastic fluids ; they exhibit elastic recovery from deformations which occur during flow. The time taken for recovery is known as the relaxation time, a purely viscous fluid having a relaxation time of zero.

For most dilute polymer solutions, such as those used in enhanced recovery, viscoelastic effects may be neglected and the stress-strain rate equation may be written as follows:

$$\vec{\vec{T}} = -p\vec{\vec{I}} + \vec{f}(\vec{\vec{D}})$$

where

$\vec{T}$ is the stress tensor (Fig. A.7.21):

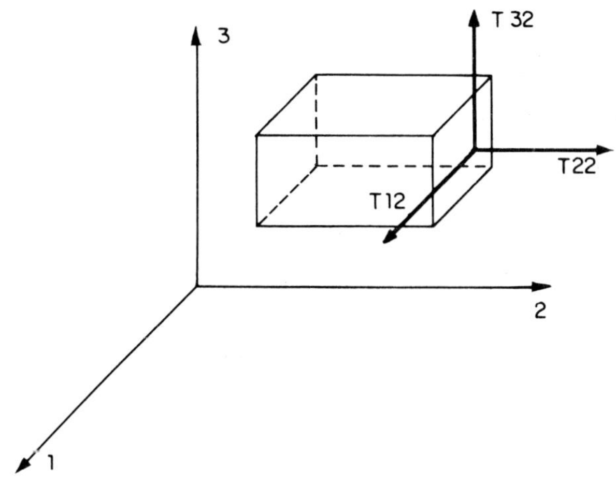

Fig. A.7.21.

$$\vec{\bar{T}} = \begin{vmatrix} T_{11} & T_{12} & T_{13} \\ T_{21} & T_{22} & T_{23} \\ T_{31} & T_{32} & T_{33} \end{vmatrix}$$

$\vec{\bar{I}}$ is the unit tensor,
$\vec{\bar{f}}$ is a tensor function,
$\vec{\bar{D}}$ is the instantaneous strain rate tensor,
p is the pressure.

2. Laminar flow

This type of flow is that studied with standard rheological instruments which measure the tangential stress ($\tau = |T_{12}| = |T_{21}|$) and, less easily, the change in normal stress as a function of the shear rate $\dot{\gamma}_c$.

a. Time dependency

It may be assumed that, since the strain rate is constant, the stresses in polymer solutions used for enhanced recovery are not time dependent. That is, that the fluids are neither thixotropic (shear stress and apparent viscosity decreasing with time), nor rheopectic (shear stress and apparent viscosity increasing with time).

b. Shear diagrams

The shear diagram for a polymer solution is shown in Fig. A.7.22.

For a Newtonian fluid of course, the shear diagram is a straight line passing through the origin (see Fig. A.7.23).

The apparent viscosity is given by $\mu = \dfrac{\mathscr{T}}{\dot{\gamma}}$ and the relationship between viscosity and shear rate for a polymer solution is shown in Fig. A.7.24.

An analytical equation for μ for each polymer solution may be obtained in the form :

$$\mu - \mu_\infty = \frac{\mu_0 - \mu_\infty}{1 + \left(\dfrac{\mathscr{T}}{\mathscr{T}_{1/2}}\right)^{\frac{1}{n} - 1}}$$

$\mathscr{T}_{1/2}$ is the stress at which the viscosity would be $\mu_0/2$.

The decrease in μ with $\dot{\gamma}$ is a characteristic of pseudoplastic behaviour.

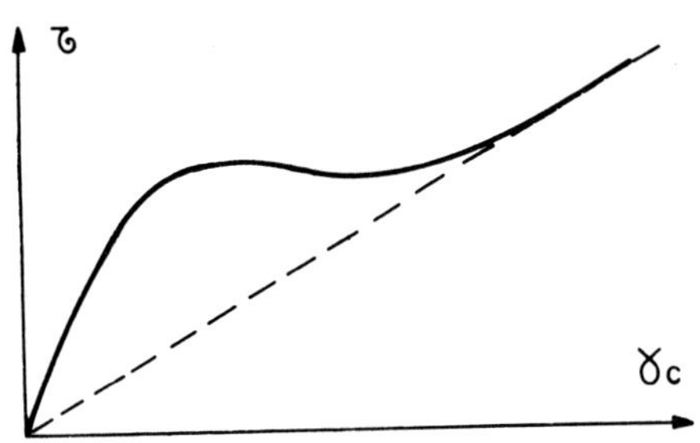

Fig. A.7.22. Shear diagram for a polymer solution.

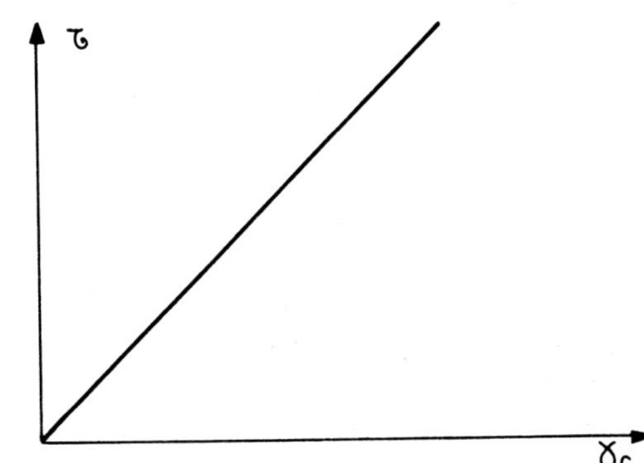

Fig. A.7.23. Shear diagram for a Newtonian fluid.

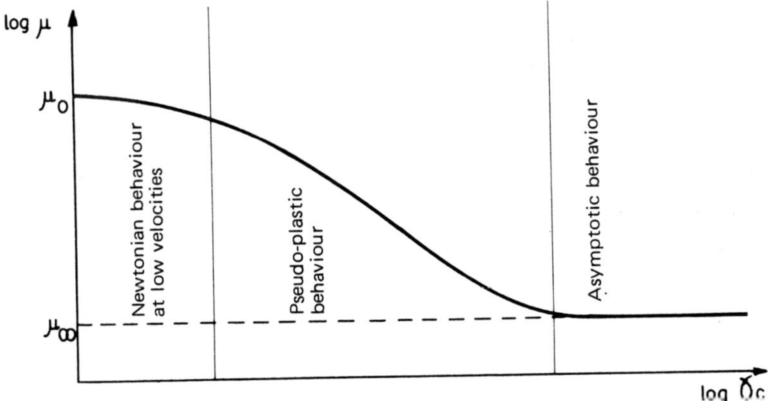

Fig. A.7.24. Plot of μ vs. γ for a polymer solution.

3. Flow with longitudinal strain

Longitudinal strain is that which causes an acceleration or deceleration of a fluid particle along its flow path. Various researchers have measured the "longitudinal viscosity" $\mu_e = \frac{T_{11}}{\dot{\gamma}_e}$ of polymer solutions. They have shown that these solutions have Trouton ratios μ_e/μ_c between 270 and 30,000 (Newtonian fluids would have a Trouton ratio of 3). The very high values of μ_e may explain the magnitude of the pressure losses which occur in porous media. If we refer to the representation of pores as a series of conic sections it is clear that the fluid particles will be successively accelerated and decelerated (see Volume I of this production course).

At present, in the quantitative application of rheological properties to enhanced recovery projects, fluids are considered to be purely viscous; only pseudoplastic effects are taken into account.

REFERENCES

General

1 PYE, D.J., Improved secondary recovery by control of water mobility. *JPT,* August 1964, p. 911-916.

2 SIMANDOUX, P., IAP Conference in 1969 (unpublished).

Polymer solutions

3 ARNAUDEAU, M., "Drainage des gisements d'huile par injection des solutions de polymères. Mise en œuvre sur champ". Rapport IFP réf. 20 586, septembre 1972.

4 CHAUVETEAU, G., DESREMAUX, L. and JACQUIN, C., "Drainage des gisements d'huile par injection des solutions de polymères". Rapport IFP 20 586, septembre 1972.

5 COLEMAN, MARKOVITZ, NOLL, *Viscosimetric Flows of non Newtonian fluids.* Springer-Verlag, 1966.

6 SANDIFORD, B.B., "Laboratory and field studies of waterflood using polymer solutions to increase oil recoveries". *JPT,* August 1964, p. 917-922.

7 SHERBORNE, J.E., SAREM, A.M. and SANDIFORD, B.B., "Flooding oil-containing formations with solutions of polymer in water" *7th World Petroleum Congress,* Mexico 1967, Proceedings, Elsevier Publishing CY, p. 509.

8 STAHL, C.D., "The use of polymers in water flooding". *Prod. Mon.* Feb., March, April 1966.

9 STAHL, C.D., "Displacement of oil by polymer flooding". *Prod. Mon.* May, June, July, 1966.

Foam injection

10 De LAMBALLERIE, G., "Etude expérimentale du drainage à la mousse". Séminaire ARTFP, juin 1967 – IFP réf. 14 587.

11 MINSSIEUX, L., "Utilisation des mousses pour la récupération du pétrole". IFP réf. 18 963, février 1961.

12 MINSSIEUX, L., "Comportement des mousses en milieu poreux". Colloque ARTEP, 7-8-9 juin 1971, IFP Rueil.

Surfactants

13 BAVIERE, M., "Drainage de l'huile par des solutions aqueuses de tensio-actifs". Rapport IFP 20 744, novembre 1972.

Micro-emulsions

14 BAVIERE, M., "Les microémulsions". *Rev. Inst. Franc. du Pétrole,* XXIX, 1, 1974, p. 41.

Carbon dioxide injection

15 MARTIN, J.W., "The use of Carbon Dioxide in Increasing the Recovery of Oil", *PM* May 1951, p. 13-15.

16 MARTIN, J.W., "Additional Oil Production Through Flooding with Carbonated Water". *PM,* July 1951, p. 18-22.

17 TORREY, P.D., "New Techniques for Improving Oil Recovery". *PM*, January 1958, p. 27-40.

18 HOLM, L.W., "Carbon Dioxide Solvent Flooding for Increased Oil Recovery". *Trans. AIME*, Vol. 216,1959, p. 225-231.

19 SIMON, R. and GRAUE, D.J., "Generalized Correlations for Predicting Solubility, Swelling and Viscosity Behavior of CO_2 – Crude Oil System". *JPT* January 1965, p. 102-106.

20 HOLM, L.W., "CO_2 Requirements in CO_2 Slug and Carbonated Water Oil Recovery Processes". *PM*, September 1963, p. 6-8, 26-28.

21 TUMASJAN, A.B., PANTELEEV, V.G. and MEJNCER, G.P., "Influence de l'anhydride carbonique sur les propriétés physiques du pétrole et de l'eau". *Nefteprom. delo*, 1969, n° 2, p. 20-23.

22 CRAWFORD, H.R., NEILL, G.H., BOCY, B.J. and CRAWFORD, P.B., "Carbon Dioxide. A Multipurpose Additive for Effective Well Stimulation". *JPT*, March 1963, p. 237-242.

23 BARDON, C., Unpublished IFP study.

24 BALINT, V., "Méthode de prévision du comportement d'un gisement d'huile balayé par du gaz carbonique". Communication n° 27, Colloque ARTEP, 7-9 juin 1971.

25 RATHMELL, J.J., STALKUP, F.I. and HASSINGER, R.C., "A laboratory investigation of miscible displacement by carbon dioxide". SPE preprint no 3 843, 3-6 October 1971.

26 BAN, A. and BALINT, V., "Résultats des essais de déplacement du pétrole par du CO_2 faits à l'échelle laboratoire et semi-industrielle et examen de l'applicabilité du procédé". *Rev. Inst. Franc. du Pétrole*, vol. XIV, 3, p. 302-336.

27 DICHARRY, R.M., PERRYMAN, T.L. and RONQUILLE, J.D., "Evaluation and design of a CO_2 Miscible Flood Project. SACROC Unit, Kelly-Snyder Field." *JPT* Nov. 1973, p. 1 309.

28 HOLM, L.W. and O'BRIEN, L.J., "Carbon Dioxide Test at the Mead-Strawn Field". *JPT* April 1971, p. 431-442.

Subject Index

ACHEVE D'IMPRIMER
EN JUILLET 1980
PAR L'IMPRIMERIE LOUIS-JEAN
05002 – GAP
N° d'éditeur 514 – N° d'impression 375-1980
Dépôt légal : 3e trimestre 1980

IMPRIME EN FRANCE